BOB HARRIS' *Guide to* STAMPED CONCRETE

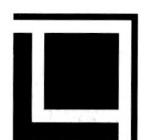

This book is dedicated to my father who taught many to go the extra mile, while mentoring me, as well as many others, in the field of concrete construction. Also, to my mom Mary Lou, for staying strong through difficult times and always being there for her family. To Lee Ann, the love of my life, for putting up with long hours and emotions, as well as all of her contributions to this book. To Melanie, Robby and Mara for being patient during insane work schedules. To the thousands of students and contractors we have had the privilege of working with that are proud to choose decorative concrete as their livelihood. Finally, to our friend Mike Speach from Tropical Toppings, for helping us out with some interesting projects when needed. Hopefully you won't have to hear the distinct sound of your trowel scraping across hard concrete, and I also hope your blisters have finally healed!

Copyright © 2004
By the **Decorative Concrete Institute, Inc.**, and **ConcreteNetwork.com, Inc.**

All rights reserved. No part of this book may be reproduced, stored in a retrieval system, or transmitted by any means—electronic, mechanical, photocopying, or otherwise—without written permission from the publisher.

Although every precaution has been taken in the preparation of this book, the publisher and the author assume no responsibility for errors or omissions. Nor is any liability assumed for damages resulting from the use of the information contained herein.

International Standard Book Number: 0-9747737-1-9

Printed in the United States of America

First Printing: September 2004

Credits

Publisher	Jim Peterson, ConcreteNetwork.com, Inc.
Associate Producer	Lee Ann Stape, Decorative Concrete Institute, Inc.
Executive Editor	Anne Balogh, ConcreteNetwork.com, Inc.
Design	Christina Wilkinson, Sabre Design & Publishing
Photography	JB Productions
Publicist/Marketing	Khara Betz, ConcreteNetwork.com, Inc.
Cover Photography	**Front Cover:** Top to bottom - QC Construction Products; QC Construction Products; Brickform Rafco Products **Back Cover:** Center photo - SuperStone, Inc.; Top to bottom - Davis Colors; Verlennich Masonry and Concrete; Concrete Impressions.

Special thanks to Jim Peterson, Khara Betz and the entire Concrete Network Staff and Anne Balogh for their devotion and dedication to these guides. It is so refreshing to work with a group that is passionate and devoted to whatever project they choose to take on.

Bob Harris' Guide to Stamped Concrete, published by the Decorative Concrete Institute, Inc. and ConcreteNetwork.com, Inc., is the second in a collection of guides for construction professionals on popular decorative concrete topics. The first guide in the collection, *Bob Harris' Guide to Stained Concrete Interior Floors* was printed in January 2004 and is currently available for purchase.

ABOUT THE AUTHOR

Bob Harris is known worldwide in the decorative concrete industry and is founder and president of the Decorative Concrete Institute in Douglasville, Georgia.

He has personally placed and/or supervised the placement of over 3 million square feet of decorative concrete, including work for some of the major Disney theme parks in Orlando, Florida.

He has conducted hands-on training seminars in architectural concrete in locations around the world, in addition to being a part of technical support, and research and development at a large decorative concrete manufacturer for almost a decade.

In addition to sharing his expertise with others through his involvement with numerous industry associations, Harris has been a presenter at four consecutive World of Concrete Trade Shows. At the 2004 World of Concrete Trade Show, Harris delivered four presentations on topics including: Acid Etch Staining, Sandblast Stencil Techniques, How to Get Started in Decorative Flatwork and Advanced Decorative Overlays. He also spoke at the 2004 World of Concrete Mexico Trade Show.

His dream for many years has been to put down onto paper the knowledge he has received from his extensive field experience and training to teach others the many techniques he has learned.

His dream was realized in early 2004 with the release of the first book in the Bob Harris Decorative Concrete Collection, *Bob Harris' Guide to Stained Concrete Interior Floors*. The book has been an industry bestseller since its release in February 2004 — selling through the American Concrete Institute, Portland Cement Association, The Concrete Network, and dozens of construction supply houses and architectural bookstores across the United States and around the world.

Bob Harris' Guide to Stamped Concrete is the second book in the Bob Harris Decorative Concrete Collection and was written with a simple goal: to provide the most useful reference available for learning and improving stamped concrete skills.

BOB HARRIS

INTRODUCTION

As more and more homeowners, builders and designers realize the flexibility and design options of working with concrete, it is clear to me, that as a decorative concrete contractor, you cannot afford to miss out on the opportunity to create quality stamped concrete. This growing demand for stamped concrete requires contractors that are skilled in stamping and coloring concrete, and who are prepared to take on unique projects. My purpose with *Bob Harris' Guide to Stamped Concrete* is to share the knowledge I have gained from my experiences to provide you with a solid foundation for producing quality stamped concrete.

Stamping concrete is a job that requires effort, timing and skill. In this guide, I have tried to clearly explain each of the many facets of stamping concrete so that you are equipped to know how to prepare concrete for stamping, when and how to place the stamps, and how to complete the work for a successful and beautiful finished product.

It is my hope that by sharing the lessons learned from my experiences you can sharpen your skills and avoid any mistakes or problems that may arise in your work. After many years of stamping concrete, I have gained an invaluable repertoire of tips and tricks that have helped streamline the stamping process, and improved many of my stamping techniques. The goal of this guide is to share those lessons learned to save you time, and give you an advantage in the process of stamping concrete.

With a complete, illustrated description of the entire stamping process, this guide provides an in-depth account of all aspects of stamping concrete. In addition, you will get solid information on grading, setting forms, and installing reinforcement. You'll also get detailed information about concrete placing and finishing, along with valuable information on important topics, including:

- Sources for stamping design ideas
- Maximizing your profits by knowing what to charge
- Concrete mix considerations for stamping concrete
- Site conditions affecting stamped concrete work and what to do to avoid problems
- How to prepare concrete for stamping, and tips for striking off and finishing concrete
- Important steps to applying color hardener

You will also get information on tools that are essential for successful stamping; important issues to avoid when stamping; fixing minor flaws in stamped concrete work; effective techniques for the application of sealers; 10 ways to promote and sell your stamped concrete work; and how to distinguish your stamped concrete work from competitors.

While only time and on-the-job experience can truly familiarize you with the stamping process, this guide can provide you with a comprehensive understanding of both the fundamentals of stamping concrete and the advanced techniques being used by seasoned pros.

L.L. GEANS CONSTRUCTION COMPANY

TABLE OF CONTENTS

	About the Author	5
	Introduction	7
1	Why Stamped Concrete Is So Popular	11
2	Stamped Concrete Offers a Wide Spectrum of Design Options	15
3	Where Do Good Designs for Stamped Concrete Come From?	18
4	Budget Analysis of Stamped Concrete	20
5	Site Conditions Affecting Stamped Concrete Work	25
6	Mix Design Considerations for Stamped Concrete	28
7	Improving the Durability of Stamped Concrete	32
8	Establishing Expectations with the Builder, Architect, and Owner	36
9	Writing a Fair Contract	40
10	Chronicling Your Work	43
11	The Importance of Safety	46
12	Stamped Concrete Pictorial Overview	51
13	Methods of Coloring Stamped Concrete	58
14	Subgrade Preparation	63
15	Erecting the Forms	64
16	Installing Reinforcement	68
17	Placing the Concrete	72
18	Striking Off and Finishing the Concrete	79
19	Applying Color Hardener	83
20	Applying the Release Agent	88
21	The Concrete Stamping Process	93
22	Curing Stamped Concrete	101
23	Installing Joints	104
24	Release Removal	110
25	Fixing Minor Flaws	115
26	Sealing Stamped Concrete	118
27	Tools, Equipment, and Supplies	122
28	How to Sell Stamped Concrete Work	128
29	Taking Stamped Concrete Over the Top	131
	Glossary	134
	Resources	138

Stamped concrete makes a dramatic impression when it complements a home's architecture. Here, a sidewalk-slate-patterned driveway in a rich gray hue enhances the gray-toned color scheme of this upscale home.

CHAPTER 1

WHY STAMPED CONCRETE IS SO POPULAR

There are many reasons why a growing legion of homeowners, businesses, and municipalities are choosing stamped concrete to enhance their landscapes and buildings.

One is permanence: Concrete is one of the most durable building materials known to man, making it ideal for pavements subject to heavy wheel or foot traffic, such as driveways and sidewalks. After final set, concrete achieves a hardness equivalent or superior to stone.

Cost is another reason why many people choose stamped concrete over alternative decorative paving materials, such as masonry paving units or natural stone. Concrete can mimic the beauty of these materials, but is much more economical. (See Chapter 4, *Budget Analysis of Stamped Concrete*, for stamped concrete cost comparisons.)

Much of this cost savings is due to concrete's speed of installation when compared with natural materials, which allows installers to be more productive. Stamped concrete can simply be poured or

Stamped concrete is an ideal surface for pool decks and patios, simulating the beauty of natural stone while permitting curvilinear borders that gracefully frame pool edges.

pumped into place; no lifting or placing of heavy individual units is required. Installing a 2,000 square foot stamped project can take five days or less from start to finish, versus 10 to 12 days for the installation of natural stone.

Probably the biggest reason for the popularity of stamped concrete is design flexibility. Unlike stone, concrete offers unlimited design options because it can be shaped, imprinted, textured, and colored to achieve almost any look imaginable. If the ultimate goal is to mimic a natural material, that's possible too with the creative blending of different colors. I often study the actual colors of the stone we are trying to replicate, and that helps me visualize the best colors to use to enhance the realism of the stamped concrete design.

The texture of stamped concrete can also be a big plus. If you are paving on a slope, for example, stamped surfaces can provide additional traction. On flat surfaces, the courser texture can help prevent slips and falls, improving the safety of public walkways.

Another attribute of stamped concrete is aesthetic value. Because of its design flexibility, stamped concrete can blend harmoniously with almost any type of architecture, whether it's a brick house or a commercial building clad in limestone. In many cases, the concrete outside a building is the first structural element people encounter. When it complements the architecture, it can make quite an impression. Retailers often use stamped concrete outside their stores to attract the attention of passersby, and thus encourage them to come in and browse.

Another feature unique to stamped concrete is the ability to customize the product. Some manufacturers produce custom stamps that allow you to imprint designs found in nature, such as a fern leaf or an animal footprint, or one-of-a-kind graphics, such as a company logo.

Where is stamped concrete being used?

Stamped concrete is especially popular in the residential market—with driveways, pool decks, and patios being in the greatest demand. Probably about 60 percent of the work I do is for this market. A lot of people are even using stamped concrete for the interior floors of their homes, to mimic slate or tile or to produce elaborate patterns.

But you're also seeing more stamped concrete outside commercial and public facilities. On a trip to Hawaii, I noticed the pavements outside the Maui airport are virtually all stamped concrete. Many ballparks, stadiums, and theme parks are also specifying stamped concrete, primarily due to its aesthetic appeal, competitive cost, and quick installation.

Even municipalities and transportation departments are starting to use stamped concrete. For example, I've seen brightly colored stamped brick designs used in concrete median islands dividing roadways to make them more visible.

Decorative stamped concrete is no longer considered a niche market. It has now hit the mainstream!

Median barriers of stamped concrete balance beauty with safety, dividing roadways while making them more scenic.

Stamped concrete design options are not limited to stone or brick patterns. With custom stamps, unique graphics, logos, and other distinctive works of art are possible.

Because of its design flexibility, stamped concrete can blend harmoniously with almost any type of architecture or landscape.

Benefits of Stamped Concrete

Permanence	Concrete is one of the most durable materials on earth. When properly installed and maintained, stamped concrete pavements will last for decades.
Competitive in cost	Stamped concrete is often more cost-effective than natural stone, slate, or masonry paving units.
Speed of installation	A stamped concrete pavement can be placed in about half the time as a natural stone pavement of equivalent size.
Design flexibility	Stamped concrete offers unlimited design options. It can be shaped, imprinted, textured, and colored to achieve almost any look imaginable.
Texture	The coarser texture of stamped concrete surfaces provides greater traction and safety.
Aesthetic value	Stamped concrete can blend harmoniously with almost any type of architecture or landscape, making it suitable for everything from residential pool decks to walkways at public airports.
Ability to customize	With custom stamps, it's possible to imprint designs found in nature, company logos, and other unique graphics.

13

Integrating stamped concrete with other landscaping elements adds interest and drama. Here, rectangular fields of textured concrete are softened by strips of grass to provide a serene setting for a pool.

CHAPTER 2

STAMPED CONCRETE OFFERS A WIDE SPECTRUM OF DESIGN OPTIONS

With countless color combinations and hundreds of patterns available, stamped concrete offers unlimited design possibilities.

In today's marketplace, you'll find a multitude of manufacturers offering a wide assortment of stamping tools (see the list of resources on page 138). Tool options include both rigid and semi-rigid polyurethane mats and flexible texturing skins. Some stamp suppliers carry hundreds of standard and custom patterns, ranging from slate, to brick, to cobblestone, to botanical and wildlife themes with the option to purchase and, in some cases, to rent.

Creative use of color adds greater realism to stamped surfaces, allowing them to mimic just about any other material. I consider stamped concrete to be a "faux finish"—something intended to replicate something else, like a brick paver, natural fieldstone, or wood planking. Such materials are rarely monochromatic. Variation in color is necessary to duplicate Mother Nature's palette.

If you look at and study rocks, for example, you'll notice distinct color variations. In concrete, we can achieve the same variation by using an integral color, maybe in the light buff range, and broadcasting dry-shake

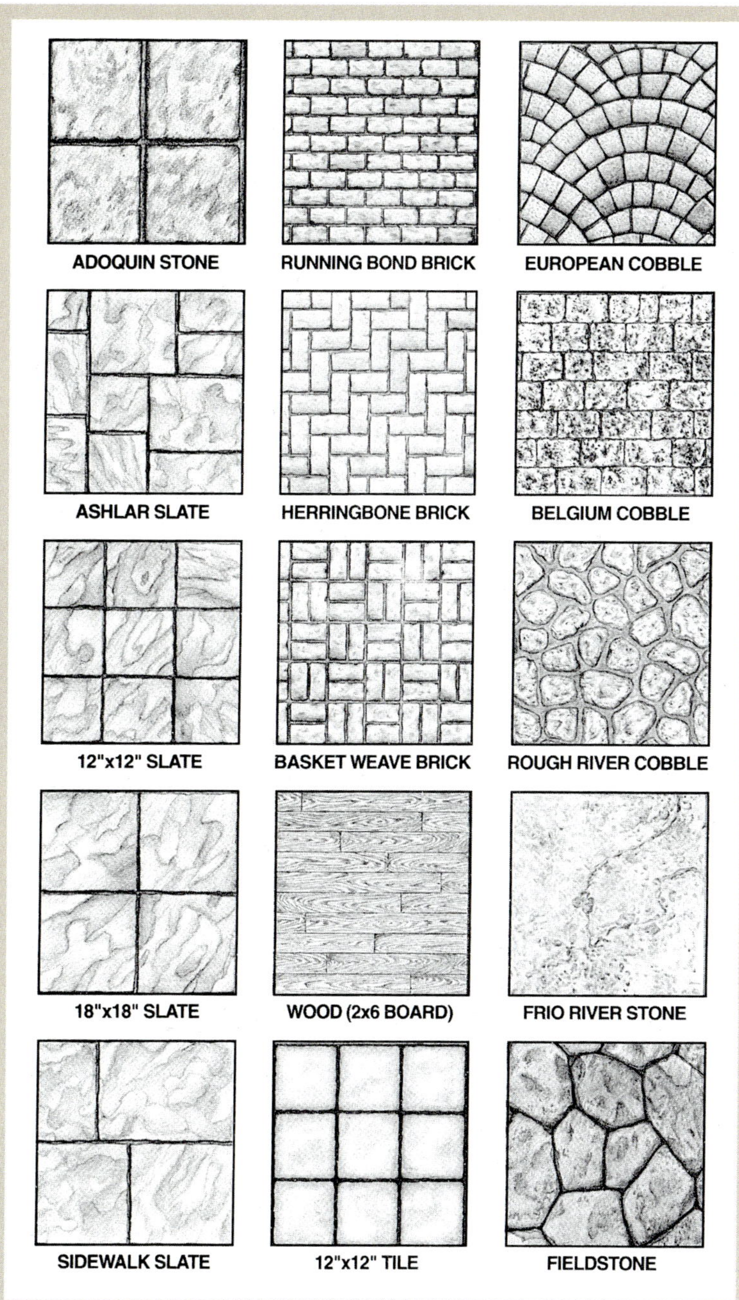

ADOQUIN STONE
RUNNING BOND BRICK
EUROPEAN COBBLE
ASHLAR SLATE
HERRINGBONE BRICK
BELGIUM COBBLE
12"x12" SLATE
BASKET WEAVE BRICK
ROUGH RIVER COBBLE
18"x18" SLATE
WOOD (2x6 BOARD)
FRIO RIVER STONE
SIDEWALK SLATE
12"x12" TILE
FIELDSTONE

CONCRETE IMPRESSIONS

Stamp suppliers carry a broad array of standard patterns.

BRICKFORM RAFCO PRODUCTS

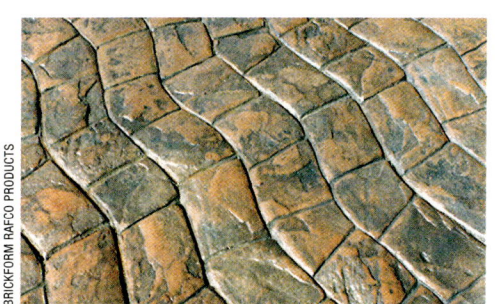

Accents, or flash colors, give stamping work greater realism. Strategically applied charcoal antiquing accents instantly add decades of weathering and wear to this rustic brick-patterned surface (left).

Evolution of Decorative Stamping

Although the popularity of decorative stamped concrete has flourished in the past decade, you may be surprised to learn that the technique has been around for more than half a century. The first stamping tools were developed in California by innovator Brad Bowman, who in the late 1940's began experimenting with ways to imprint brick, stone, and tile patterns in fresh concrete as a cost-saving alternative to hand-laid stone. He made his first tools of wood and then switched to large cast-aluminum platform stamps that looked like giant cookie cutters with handles attached. These tools made basic rectangular brick and tile impressions in concrete, leaving a joint line about 1/4 to 3/4 inch in depth between each pattern unit. His invention led to the birth of what is today the largest segment of the decorative concrete industry. Bomanite Corporation was formed in 1970 to franchise his process. However, many independent contractors have also adopted and expanded upon his techniques.

Bowman passed away in December 2000, but his original concept has continued to evolve. Although cast-aluminum platform stamps are still available, they have largely been superseded by polyurethane mats that simulate both the pattern and texture of stone, brick, slate, wood planks, seashells, and many other materials. More recently, thinner, highly flexible texturing skins have been introduced to allow imprinting of seamless texture in concrete. Custom stamps are also becoming popular for imprinting of graphic designs, such as company logos.

Another trend is the use of creative coloring techniques to give stamped surfaces more variation and drama. Dry-shake color hardeners and colored release agents are available in an increasing array of hues. With the ultimate goal of producing an authentic look, more and more contractors are using chemical stains to add accents of color or to give surfaces an antiqued, timeworn patina. Today it is common for contractors to incorporate multiple patterns, textures, and colors into their decorative stamping projects, rather than settle for a one-pattern, one-color surface. Mr. Bowman, himself, would be impressed by their creativity.

The first stamping tools, invented by Brad Bowman more than 50 years ago, were limited to basic brick and tile patterns. Today, stamping contractors can choose from a multitude of patterns and color choices.

hardener on the surface in as many as four or five different accent colors (see Chapter 13, *Methods of Coloring Stamped Concrete*). In the stamping world we call that technique "flashing." The accents, or flash colors, give your stamping work much more realism than is possible with just one solid color.

You can also achieve color contrast in other ways, which I'll describe in detail later. These include using different hues of powdered or liquid release agents or applying chemical stains.

Although stamped surfaces look great on their own, you can heighten the drama by enhancing or complementing the colors and textures with other types of finishes. For example, combine stamping and exposed aggregate to produce eye-catching borders and insets. Or place the stamped concrete in bands or fields separated by plain or broom-finished concrete. You can even combine different patterns of stamped concrete on the same project, such as a stone-patterned area with a stamped brick border. For more ideas and inspiration, read Chapter 3, *Where Do Good Designs for Stamped Concrete Come From?*

QC CONSTRUCTION PRODUCTS

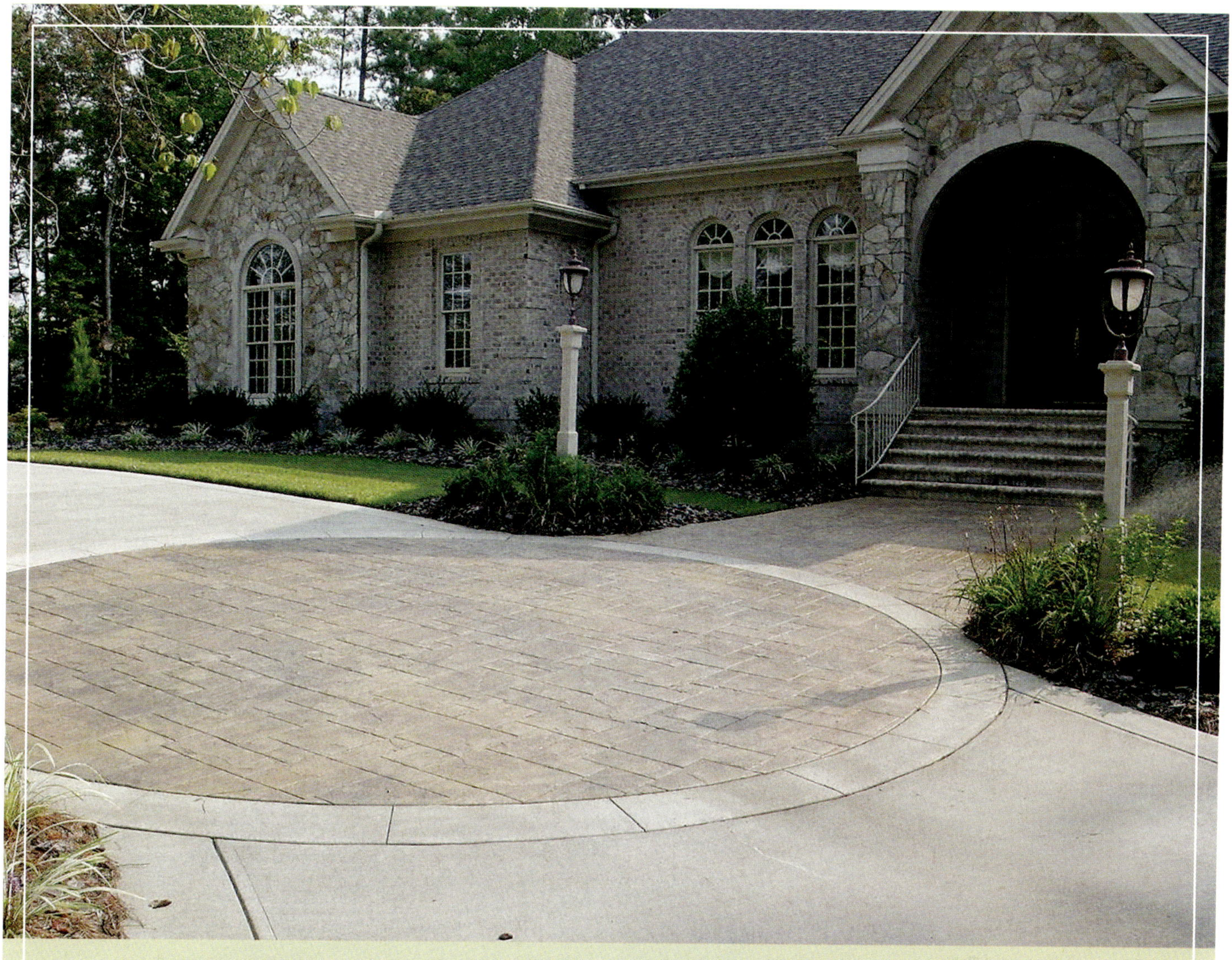

Stone-patterned concrete encircled by a band of plain concrete becomes the focal point of this curved driveway. The pattern is repeated in the sidewalk and steps leading to the home's arched entryway.

CONCRETE IMPRESSIONS

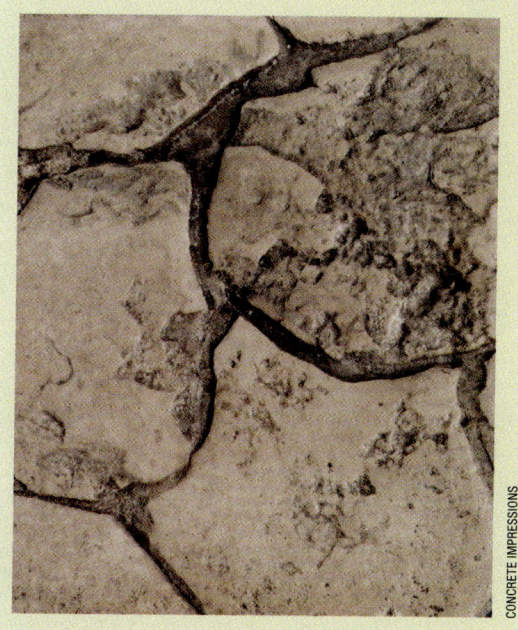

Is it fieldstone or stamped concrete? Today's stamping tools are often molded from real materials to duplicate their textures down to the finest detail.

CHAPTER 3

WHERE DO GOOD DESIGNS FOR STAMPED CONCRETE COME FROM?

You don't need to be an architect or have an artist's eye for color to be creative with stamped concrete. Coming up with good designs can be very simple. Just let the project's surrounding landscape and style of architecture inspire you.

Start by looking at the basic construction of the home or building the stamped concrete will be accenting. Is it a traditional brick house? Then the obvious choice would be to echo that brick theme in the hardscape, whether in a brick-patterned border or entire driveway. Choosing appropriate color combinations can be equally simple. The goal is to achieve harmony rather than glaring contrasts. Study the colors of the building's structural elements, such as the roof, the siding, and the trim around the windows. Incorporate these hues or complementary tones into your stamping work.

On larger projects, a lot of your designs will come from partnering and working with the designer, whether a landscaper or architect. Most likely these professionals will give you a specific set of design criteria and conceptual

Great design ideas for decorative stamped concrete may be right underfoot. You'll often encounter the handiwork of other contractors as you stroll through shopping malls, theme parks, hotel facilities, and other public spaces.

drawings you can work from.

Conversely, there will be some projects where the client will give you total design freedom. Clients often welcome creative ideas and even expect them. For these projects, you may need additional inspiration.

Some of the best sources for ideas are the brochures and literature from stamping tool manufacturers. Most major manufacturers will have pictures showing beautiful examples of projects utilizing their stamps.

Of course, this book is another excellent source for ideas. You'll see countless examples of striking stamp designs and color combinations throughout, especially in Chapter 12, *Stamped Concrete Pictorial Overview*.

Quite frankly, I've learned a lot about stamping techniques and good design from observing the work of other concrete craftsmen. If you know of other stamping contractors in your area, go out and look at their installations. In most cases, you'll find this to be a "sharing" community of professionals, where most contractors are more than willing to let you come out and admire their handiwork. Or search the Internet, where you'll find hundreds of concrete contractors displaying photos of their best decorative stamping work on their web sites.

You can also learn techniques and get design ideas from organizations such as the Decorative Concrete Council (a committee of the American Society of Concrete Contractors) and from networking at trade shows such as the World of Concrete.

Even consider looking to the work of other trades for design possibilities. The stone and marble industries, for example, often publish brochures showing the beautiful work of stone craftsmen. Try duplicating the same patterns, colors, and textures in your designs.

If you're still at a loss for ideas, take a day off with your family and visit a theme park. Whether it's Disney World, Sea World, or Six Flags, theme parks often use stamped concrete in some form or fashion—and usually quite creatively. You may even spot other people in our trade looking down and taking pictures of the concrete rather than enjoying the other amenities in the park!

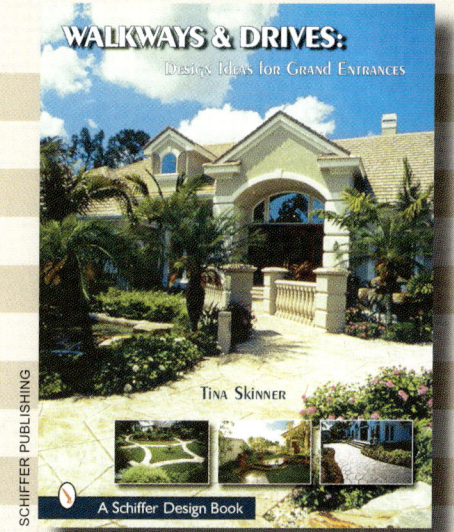 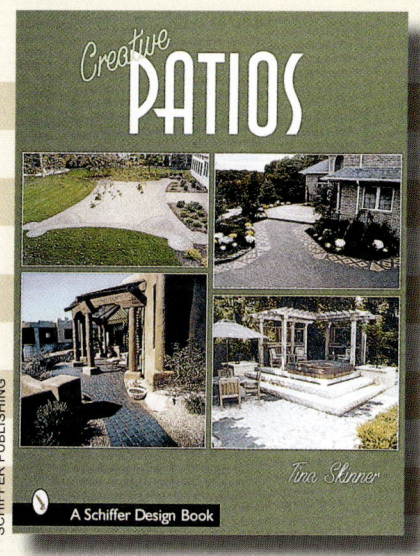

Jumpstart your creativity by browsing through design books on exterior hardscaping. Try duplicating the same patterns, colors, and textures in your own work.

 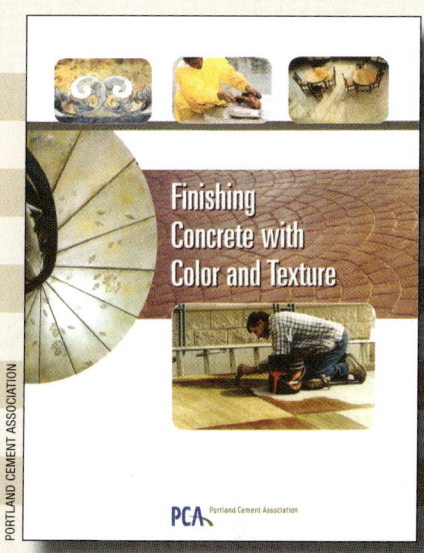

Investigate the wide selection of decorative concrete publications available from the Portland Cement Association and other industry organizations. You'll find invaluable how-to advice as well as photos showing techniques and design possibilities.

Sources for Good Design Ideas

- The architectural style of the buildings on the jobsite
- Brochures and other literature from stamp manufacturers
- The work of other decorative concrete contractors
- The work of other trades, such as stone craftsmen
- Trade shows displaying the latest products for decorative concrete
- Industry groups, such as the Decorative Concrete Council
- Theme parks
- Books on decorative concrete
- The Concrete Network web site (www.concretenetwork.com)

CHAPTER 4

BUDGET ANALYSIS OF STAMPED CONCRETE

While prices for materials and labor in your local market will influence the final cost of a stamped concrete installation, it is helpful to compare stamped concrete with alternative paving options. As the table on page 21 shows, based on the cost for a typical residential driveway, stamped concrete is an excellent choice in terms of aesthetic value and performance.

Landscapers and architects often choose stamped concrete for its unparalleled beauty, performance, and versatility. Stamped concrete has a very durable surface ideal for exterior applications subject to wear and tear, such as pool decks and driveways. It also retains its beauty for years without the color fading or wearing away.

The upfront cost of a project is only one component of the total cost equation: Maintenance expenses incurred over the life of the pavement must also be taken into account. Concrete pavements of all types generally last much longer and require less maintenance than most other paving materials. And while customers typically end up paying more for stamped concrete, particularly if they want an elaborate design, they will enjoy a magnificent-looking pavement that will last for decades with minimal maintenance. With all that going for stamped concrete, you can be confident that the price of your work will be competitive. In fact, stamped concrete is often preferred by landscape architects and designers (see Why Designers Choose Stamped Concrete on page 22).

Driveway Material	Installation Cost	Maintenance Requirements
Asphalt	$2 to $2.50 per square foot (includes 2 inches of rock base)	Seal coat required approximately every 5 years.
Concrete (uncolored, broom finished)	$2.50 to $5 per square foot	Removal of dirt or stains when necessary (by scrubbing or pressure washer).
Pavers	**Basic Design** – $4 to $6 per square foot **Elaborate Design** – $7 to $20 per square foot, depending on design features included (such as borders and multiple color and pattern combinations)	Removal of weeds and other vegetation that grow up through the joints. Erosion can also be a problem, if the pavers are installed over regular compacted fill.*
Stamped Concrete	**Basic** (use of an integral color or one color of dry-shake hardener) – $6 to $8 per square foot **Intermediate** (use of two or three colors and maybe a border in a contrasting color) – $8 to $13 per square foot **Advanced** (may include intricate borders, sawcut designs, and chemical stain accents applied by hand) – $15 to $20 per square foot	Cleaning and resealing every few years or as needed to protect the surface and maintain color vibrancy. (For procedures, see Chapter 26, *Sealing Stamped Concrete*.)
Natural stone	$25 and up per square foot	Removal of weeds and other vegetation that grow up through the joints. Some stones may require replacement after several years due to breakage or discoloration.

— MARKETS SURVEYED: SEATTLE, PHOENIX, AND ATLANTA

*A high percentage of paver installations that I've seen in the United States have been installed on a compacted granular base or sand fill. The price for pavers goes up substantially when installed on a concrete base, which is the best way to go to prevent erosion and reduce weed growth.

The cost of stamped concrete increases when more sophisticated coloring techniques are used, such as hand-applied chemical stain accents. Yet the magnificent results are well worth the investment.

Why Designers Choose Stamped Concrete

Compared with alternative decorative paving materials, such as natural stone and brick or concrete pavers, stamped concrete offers the perfect combination of aesthetics, longevity, and versatility, say many of the architects and landscape designers that I've worked with. They often choose stamped concrete for pool decks, driveways, patios, and sidewalks, for both residential and commercial projects. Here are some of the reasons why:

Wide selection of colors, patterns, and textures

"Stamped concrete allows the homeowner or designer to choose from a diversity of color, patterns, and textures that can match natural materials in some cases, and at a lower price tag," says Sean Murphy of Amenity Architects, a firm specializing in mid-range to high-end residential work. He often opts for stamped concrete instead of paving stone for pool decks and driveways.

Alex Paulson of Randall-Paulson Architects has used stamped concrete for retail and office projects throughout the Southeast, primarily to give exterior building pavements a distinctive look. "You can achieve very natural looks with it and be very innovative," he says, noting that the techniques and options for stamped concrete have come a long way in the past few years.

Less labor-intensive to install

"It's a very affordable, practical alternative," says designer Larry Grams, who specializes in residential work. "I can stamp concrete cheaper than I can install pavers. Any time you have to install individual paving units—whether brick, stone, or concrete—you have a labor component to cut and place them as opposed to just pouring concrete and applying a texture."

This labor savings often makes stamped concrete the most economical option. "For some jobs, the cost of doing stamped concrete is much less than the cost of installing natural stone or brick," says David Bennett of Bennett Landscape and Design. "Stamped concrete can be placed in two or three days versus several weeks for stone or brick." He also says that concrete is great for hard-to-access jobsites with difficult layouts. "You can pump the concrete where you need it, if necessary."

Longevity and performance

Murphy finds that stamped concrete holds up well in exterior applications subject to extensive wear and tear, such as pool decks and driveways. "Stamped concrete has a very durable surface. The color hardener makes it stronger and more resistant to scratching and water penetration," he says.

Stamped concrete, especially when integrally colored, also maintains its beauty for years without the color fading or wearing away. "The color is typically within the concrete and continuous. It develops a certain amount of patina with age that I think enhances the look," says Grams.

Ease of maintenance

For the home or building owner, stamped concrete offers the additional benefit of being easy to maintain. Generally, you can "seal it and forget it," says Murphy. "If the concrete is subject to a lot of use, it can be resealed after a few years."

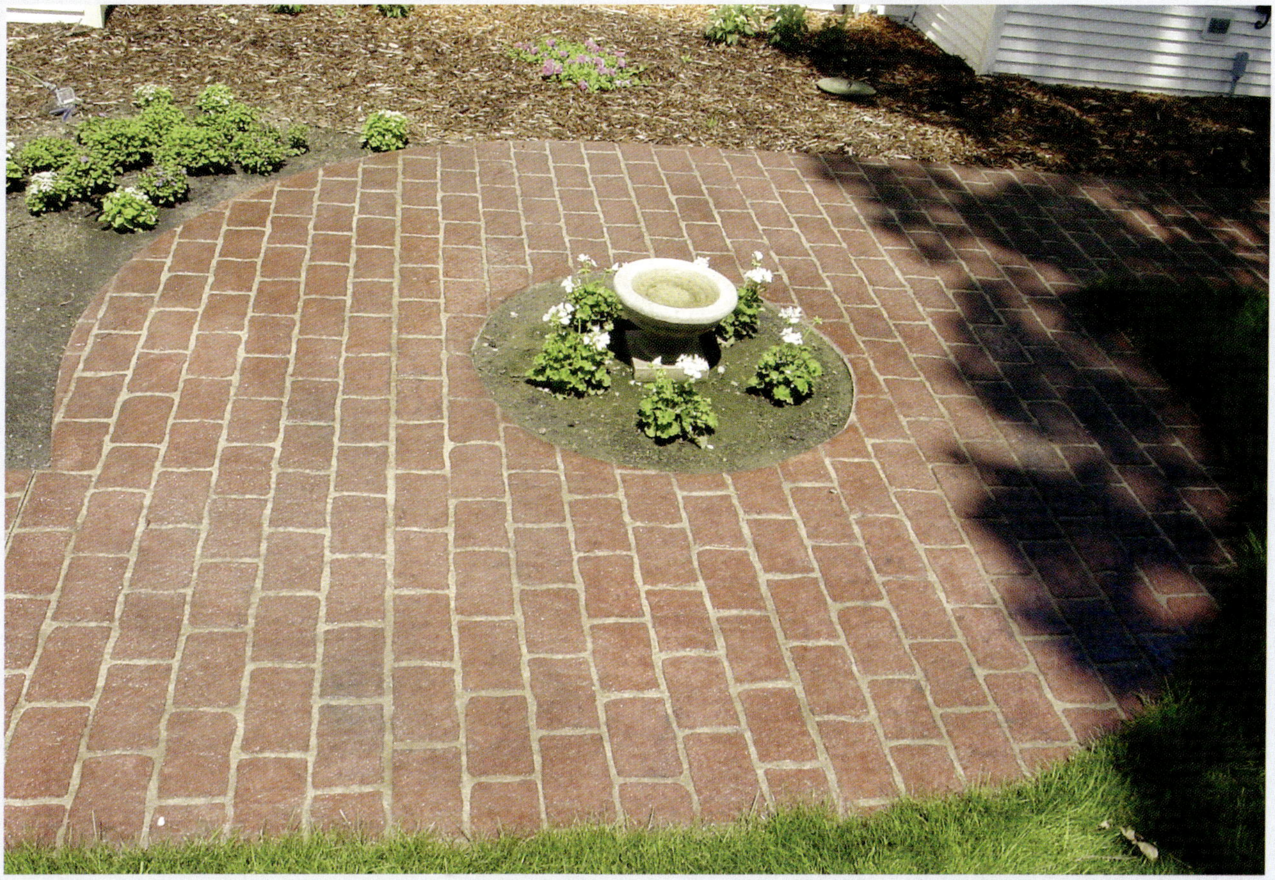

VERLENNICH MASONRY AND CONCRETE

Basic stamped concrete—using an integral color or one color of dry-shake hardener—runs about $6 to $8 per square foot, or a few dollars more per square foot than plain, uncolored concrete.

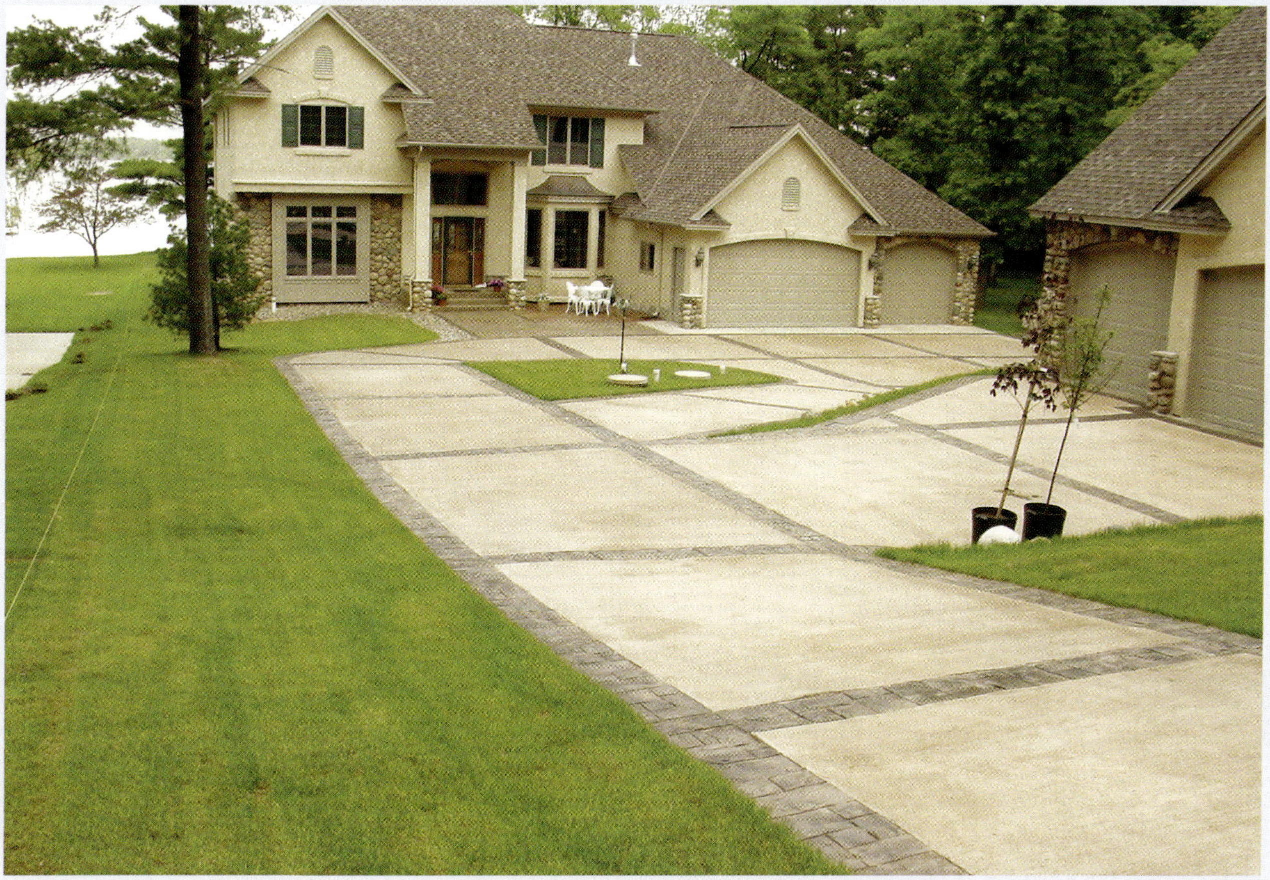

VERLENNICH MASONRY AND CONCRETE

On this project, the homeowner saved costs by using only borders and bands of stamped concrete, rather than imprinting the entire driveway. However, the overall effect is no less impressive.

DECORATIVE CONCRETE INSTITUTE

CHAPTER 5

SITE CONDITIONS AFFECTING STAMPED CONCRETE WORK

You've selected a stamping pattern, decided on the coloring techniques to use, and have all the equipment and tools necessary to complete your project (see Chapter 27, *Tools, Equipment, and Supplies*). But don't be too eager to get started. First, you need to carefully assess the worksite and address each of the issues discussed in this chapter.

Accessibility

One of the most important considerations is gaining accessibility to the site, particularly for the concrete truck. Is there a wide enough area for the truck to turn around? Is the truck able to pull into the placement area to deposit the concrete right from the chute (called "tailgating")? If not, you will need to make other provisions for transporting the concrete (see Chapter 17, *Placing the Concrete*).

You must also plan for the access of other delivery trucks and construction equipment. For example, if you have a large pour requiring a lot of excavation work to remove dirt and replace it with aggregate fill, make sure you have an access point for excavation equipment, dump trucks, and aggregate delivery trucks. Planning for accessibility can make the difference between a project that moves forward without a hitch to one that stops dead in its tracks.

Subgrade

One of the most commonly overlooked aspects of good concrete construction is starting with a properly prepared subgrade. Though you don't see the subgrade after the concrete is placed, it can make a big difference in the overall performance—and appearance—of the final product. A well-draining, compacted fill acts as a filtration system and prevents concrete settlement and subsoil erosion. And when uniformly graded, it will help to ensure a uniform thickness of concrete. If the concrete placement is thicker in some areas than others, you can actually promote plastic shrinkage cracking because the concrete is not setting at the same rate. I consider this aspect of the work to be so important that I've devoted an entire chapter to it (see Chapter 14, *Subgrade Preparation*).

Frozen ground

Concrete should not come into contact with any frozen surfaces, such as a frozen subgrade, ice, or snow. Concrete poured on a frozen subgrade can crack or settle unevenly when the subgrade thaws. A frozen subgrade also cools the bottom of the concrete slab when it's placed, which retards setting and early strength gain.

Ideally, the jobsite will permit easy access for the concrete truck, enabling the vehicle to back right up to the placement area to deposit its load directly from the chute. If not, you'll need to make other provisions to transport the concrete, such as wheelbarrows or a power buggy.

Contractors in northern climates have various methods for thawing out frozen ground, depending on the depth of the frozen surface. These include the use of insulated blankets and hydronic ground heaters, which circulate a heated fluid through flexible tubing. Check with your local ready-mix producer for the best practices in your area.

Drainage

Your customers won't be happy if standing water lingers on their stamped pavements after every rain shower. For most stamped flatwork on fairly level ground, I've found that an 1/8 inch slope per foot (or 1 inch per 8 feet) is adequate to ensure water runoff and prevent ponding. But patterns with heavy textures may require more slope than that. For example, a pattern with crevices more than 3/8 inch deep will readily hold water if you don't have sufficient slope. Typically, in this instance, a 1/4 inch per foot or 2 percent slope would be more appropriate. If elevation constraints prevent you from getting the slope you need, use a less-textured stamp design, with a pattern depth of maybe 1/4 inch or less. If the site is in an area of extremely poor drainage, you may need to regrade it or install swales or drains to prevent ponding of water.

Interference from other trades

If you're on a large project site where other trades are at work, it's important

The subgrade for stamped concrete work should be properly graded to ensure a uniform concrete thickness. If the concrete placement is thicker in some areas than others, inconsistent setting will occur and in some cases, plastic shrinkage cracking can result.

to establish your placement schedule and work boundaries early on so other crews don't unintentionally ruin your efforts. If your crew is trying to establish grade elevations, for example, and another crew's truck drives over the grade stakes, you'll end up with angry workers and a delayed job. To minimize such interference, determine your construction schedule in advance and share that information with other trades (see Chapter 8, *Establishing Expectations with the Builder, Architect, and Owner*). In a perfect world, nobody else would be allowed on the jobsite during the time you need to finish the stamped concrete work, from start through completion. But in most cases, that just won't be possible and communication and compromise will be necessary.

Protecting the work area with barricades or signs—before, during, and after the pour—is extremely critical. During the construction phase, let's say you set the forms and then get called out to another job. Two days later, you come back and find that someone has bumped the forms with a wheelbarrow and knocked them out of alignment. Obviously, if your forms are out of whack, the concrete placement will be too.

After stamping is complete, you still need to barricade the work area to keep other trades from tracking dirt or dragging tools and equipment across the freshly placed and colored concrete. Barricades and proper signage are especially important on public-works projects. I've had kids ride their bicycles and people walk through our freshly stamped concrete. In one case, a woman drove her car through a crosswalk we had just completed. Be sure to protect your work during all stages of the project!

Project layout

What's the layout of the worksite? In addition to concerns about access and drainage, you may have to contend with other site challenges. For example, we did a multilevel stamped pool deck in Florida that had four different elevation changes. The owner wanted the pattern to start at one end of the deck, go all the way up the different tiers, and then come back down to the starting point, with all the pattern lines matching perfectly. That's almost impossible to do without precisely delineating starting and stopping points for the stamp pattern. Stamps have a tendency to drift, so if you don't continually check the positioning with a string line or reference points (such as marking stamp positions on the edge of the form), it's virtually impossible to start at one end of a deck and come all the way around to the opposite end and have the pattern match up, even for the most skilled stamper. So plan your stamping layout in advance and, if necessary, do a dry run with the stamping mats to ensure success.

Use barricades to keep homeowners or other trades from disturbing the work area or treading on the freshly stamped concrete.

CHAPTER 6

MIX DESIGN CONSIDERATIONS FOR STAMPED CONCRETE

Concrete isn't a generic product that's exactly the same every time you order it. There will always be variations, however slight. Sometimes those variations are intentional, so you can tailor the concrete to meet specific job needs.

Most professional stamping contractors have several mixes they like to use under different circumstances, often based on weather conditions and the size and time frame of the project. They may have a summer mix and a winter mix, or even a floor mix for interior jobs. But don't be too concerned about the mixes preferred by other contractors. The best mix is the one that works in your area for your purposes—and gives you predictable results.

Before getting into some of the mechanics of concrete mix design, I must emphasize the importance of partnering with your ready-mix supplier. When you have a good rapport with the plant manager and the truck drivers, they will be more willing to work with you if a mix is giving you problems, such as setting too fast or too slow. Slight adjustments to a mix design can often make it more workable and give you better results.

Choosing a basic mix design

Your first step is to find a basic mix design that is reliable and meets the majority

The best concrete mix is the one that works in your area for your purposes—and gives you predictable results.

of your needs. Again, your ready-mix producer can help here, but often it takes a bit of trial and error to discover what works best. An excellent reference I always recommend to my students is the Portland Cement Association's book, *Design and Control of Concrete Mixtures* (see For More Information on Concrete Mixtures on page 31). Note that on large commercial or municipal projects, the architect/engineer will typically specify the minimum requirements for the concrete mix design, in terms of compressive strength, water-cement ratio, percentage of air entrainment, and other performance factors.

For stamping work, you typically want a "fatty" mix—one with a sufficient amount of cement paste to provide a layer of cream on the surface that will take an imprint well. Generally, I like to use a minimum of a 5 1/2 to 6 sack mix (5 1/2 to 6 bags of cement per cubic yard of concrete). I've seen contractors run into problems when they cut corners and go down to a 4 1/2 or 5 sack mix that's deficient in fines. Such a

For stamping work, a mix with a sufficient amount of cement paste often results in a better imprint.

Combating Efflorescence

Most of us have seen concrete or masonry that has been discolored by efflorescence—powdery white blotches that form when soluble salts in the concrete are carried to the surface and left there to harden. On colored concrete, especially darker tones, these white deposits are particularly conspicuous and can ruin an otherwise perfect stamping job. They often show up near joints, where water passes through the concrete. Although efflorescence doesn't affect concrete durability, it's an eyesore and can be very difficult to remove. The best approach is to prevent the problem from occurring in the first place.

Here are some measures you can take to prevent efflorescence:

- Use concrete with a low water-cement ratio.

- Cure the concrete thoroughly to reduce its permeability.

- Apply a water - or solvent-based cure and seal to the concrete as early as possible (see Chapter 22, *Curing Stamped Concrete*).

- Make sure water drains away from the slab and doesn't pond on the surface.

- Tell the owner to avoid the use of deicing salts.

How can you remove efflorescence? It's best to go after the salt deposits as soon as they appear, while they are still water-soluble and can be washed away with plain water or a mild detergent. If the salt deposits have already hardened, try removing them with a diluted muriatic acid solution and a nylon scrub brush. Some manufacturers sell prepackaged solutions for efflorescence removal.

To the Rescue: Prepackaged Admixtures

On most decorative stamped concrete projects, especially larger jobs, time is of the essence. Your crew could have a limited time frame to place, float, color, and stamp the fresh concrete before it begins to set. But what happens if unexpected delays occur, due to bad weather, equipment breakdowns, or other factors? Or even worse, what if the concrete that arrives at the jobsite is already too stiff to place and finish properly or the air content is too low?

Prepackaged admixtures that you can dose on the jobsite will be an important part of your tool kit. They can give you greater control over the concrete you receive and the amount of time available for decorative stamping. Fritz-Pak Corporation (www.fritzpak.com) offers a variety of powdered admixtures conveniently packaged in water-soluble bags that can simply be tossed into a batch of concrete whenever needed. The premeasured quick-fix powders provide first aid for a number of common problems faced by stamping contractors. For example, if you have a very large area to cover and you are concerned about the concrete setting too quickly, you can use a prepackaged retarder to keep the concrete workable longer.

In addition to retarders, prepackaged admixtures available from Fritz-Pak Corporation include an air entrainer, a superplasticizer (or water reducer) that improves the workability of concrete without increasing the water-cement ratio, and a finishing aid that enhances surface finishing characteristics and makes stamping and the addition of color hardeners easier.

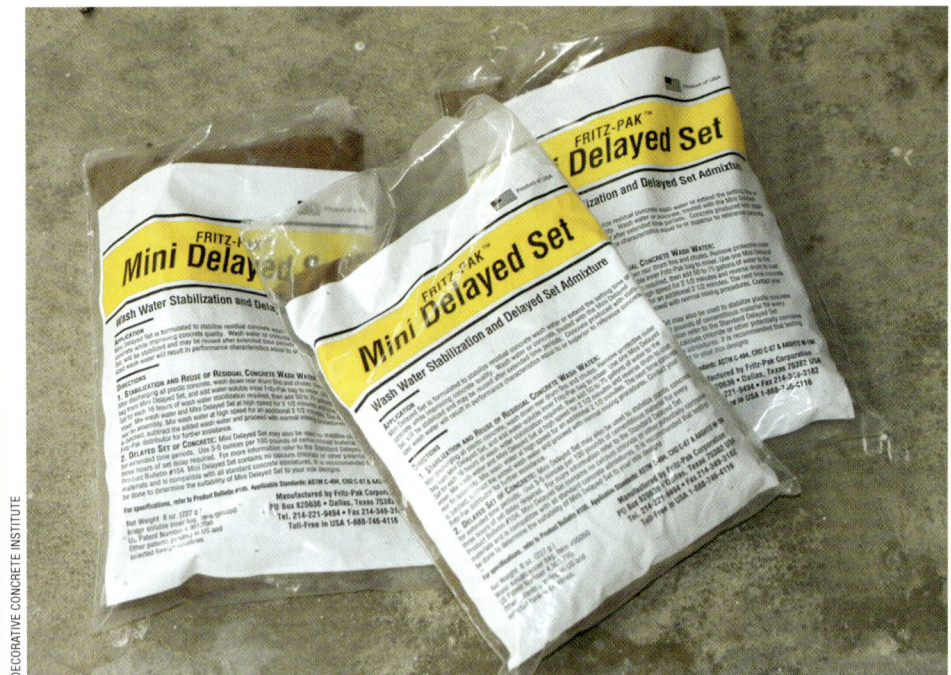

Admixtures can help you fine-tune the mix and enhance concrete performance, both in the plastic and hardened states. They can be added at the ready-mix plant or at the jobsite as the need arises.

Ready-to-use admixtures packaged in water-soluble bags can simply be tossed into a batch of concrete whenever needed. Bags of retarder are especially useful to have on hand, allowing contractors to extend the time available for decorative stamping.

lean mix won't give you the paste needed to provide a clean, well-defined imprint—and it could jeopardize the structural integrity of the concrete.

The size of the aggregate is important too. The larger the nominal aggregate size, the less the concrete will shrink. I try to keep the maximum aggregate size to about 3/4 inch. However, if you're using a pattern with an extremely deep reveal, you may need to go down to a 1/2 inch top aggregate size. If possible, avoid using a straight pea rock mix (3/8 inch aggregate), especially if the concrete is 5 or 6 inches thick. Pea rock is often used when a small-line pump is needed to get the concrete

For More Information on Concrete Mixtures

One of the most-used references on my bookshelf is the Portland Cement Association's *Design and Control of Concrete Mixtures*. This definitive volume on concrete technology, now in its 14th edition, is one I've turned to often for answers to questions about the performance and behavior of freshly mixed and hardened concrete. At 358 pages, it's packed with useful information. It discusses all the materials that go into concrete, such as portland cements, supplementary cementing materials, aggregates, admixtures, and fibers. The effects of air-entraining agents, water reducers, retarders, and accelerators are also covered. Construction topics include procedures for placing, consolidating, finishing, and curing concrete, and the precautions to take when placing concrete during hot or cold weather. You can also learn more about the commonly used tests for quality control of concrete, including tests for slump, air content, temperature, and unit weight of fresh concrete, plus strength, durability, and permeability of hardened concrete.

You can order a copy of the book (now also available on CD-ROM) at PCA's online bookstore (www.cement.org/bookstore) and at Amazon.com. Or call PCA at 800-868-6733.

to the point of deposit (see discussion on pumps in Chapter 17). In this situation, try to use a line pump that can accommodate a larger aggregate size.

Using admixtures to enhance performance

Once you get a basic mix design down and your skill level increases, explore the use of different admixtures. These products can help you fine-tune the mix and enhance concrete performance, both in the plastic and hardened states.

When used in the right proportions, pozzolans and other cement replacements (such as fly ash and granulated blast-furnace slag) can have positive effects on concrete by improving finishability, reducing permeability, reducing efflorescence (see page 29), and minimizing bleeding of color. Pozzolans can also have an effect on the setting time of the concrete. Granulated blast-furnace slag, for example, can slow setting and extend the working time of the concrete. Based on my experience, however, this retardation effect occurs only at cooler temperatures. When the air temperature exceeds 80º to 85º F, I haven't seen any retardation with granulated slag.

Keep in mind that concrete is influenced by temperature: The warmer the weather, the faster the concrete sets. To slow the set of the concrete in hot weather, try using a retarding admixture, which can be predosed at the batch plant or added by you at the jobsite (see To the Rescue: Prepackaged Admixtures on page 30). A retarder can buy you additional stamping time, especially when outdoor air temperatures heat up.

Conversely, if you are faced with cool conditions, you can use an accelerating admixture to speed the setting time of the concrete and minimize bleeding and segregation. Avoid using products that contain calcium chloride or added chloride ions, which can result in discoloration of colored concrete.

To improve the durability of concrete without sacrificing workability (or how easily the fresh concrete can be placed and finished), consider using a water reducer. These products are designed to lower water requirements by about 10 percent, greatly improving concrete strength at all ages and reducing permeability and cracking. Another admixture that contributes to durability, and one that is essential for concrete exposed to freezing and thawing conditions, is an air-entraining agent. See Chapter 7, *Improving the Durabilty of Stamped Concrete*, for more details on enhancing durability.

CHAPTER 7

IMPROVING THE DURABILITY OF STAMPED CONCRETE

In Chapter 6, we focused mainly on the characteristics of the concrete mix that make it easier to place, finish, and stamp. But it's equally important to use a mix that will result in a durable final product—one that can endure all anticipated exposure conditions.

In northern climates, exterior stamped concrete is subject to the inevitable cycles of freezing and thawing in winter. When moisture in concrete freezes, it expands and creates internal pressures that—unless relieved—can ruin your skillful stamping work. Similarly, concrete that contains too much water in proportion to the cement (called the water-cement ratio) is more likely to experience shrinkage and cracking. Here are some ways to boost the performance of your concrete and make it more resistant to damage.

Protecting concrete from freeze-thaw damage

The consequences of freeze-thaw damage can be particularly devastating to the appearance of stamped concrete. They include ugly scaling (or flaking of the surface), popouts, and cracking, all of which may continue to worsen with successive freeze-thaw cycles.

Fortunately, concrete mixes can be designed to resist freeze-thaw damage, allowing stamped concrete to hold up well

32

for years in any climate. Here are some strategies to take:

- Tell your ready-mix producer to add an air-entraining admixture to the concrete. These admixtures fill the concrete with billions of microscopic air bubbles, which help relieve the internal pressures caused by the expansion of water when it freezes. Ask for an air content of about 5 to 8 percent of the volume of concrete (or as recommended by your ready-mix producer for the exposure conditions in your area).
- Make sure the concrete contains frost-resistant aggregate. Some aggregates are porous and more susceptible to fracturing during freeze-thaw conditions, leading to popouts.

BRICKFORM RAFCO PRODUCTS

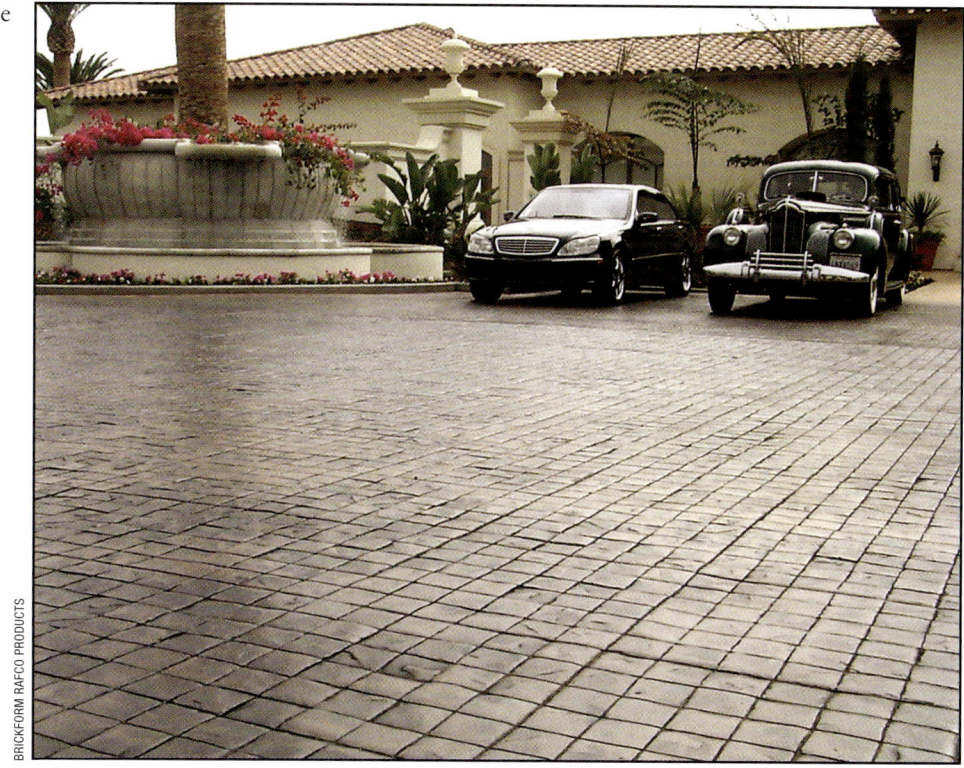

With the proper concrete mix, stamped concrete is durable enough to be used anywhere.

SUPER-KRETE INTERNATIONAL

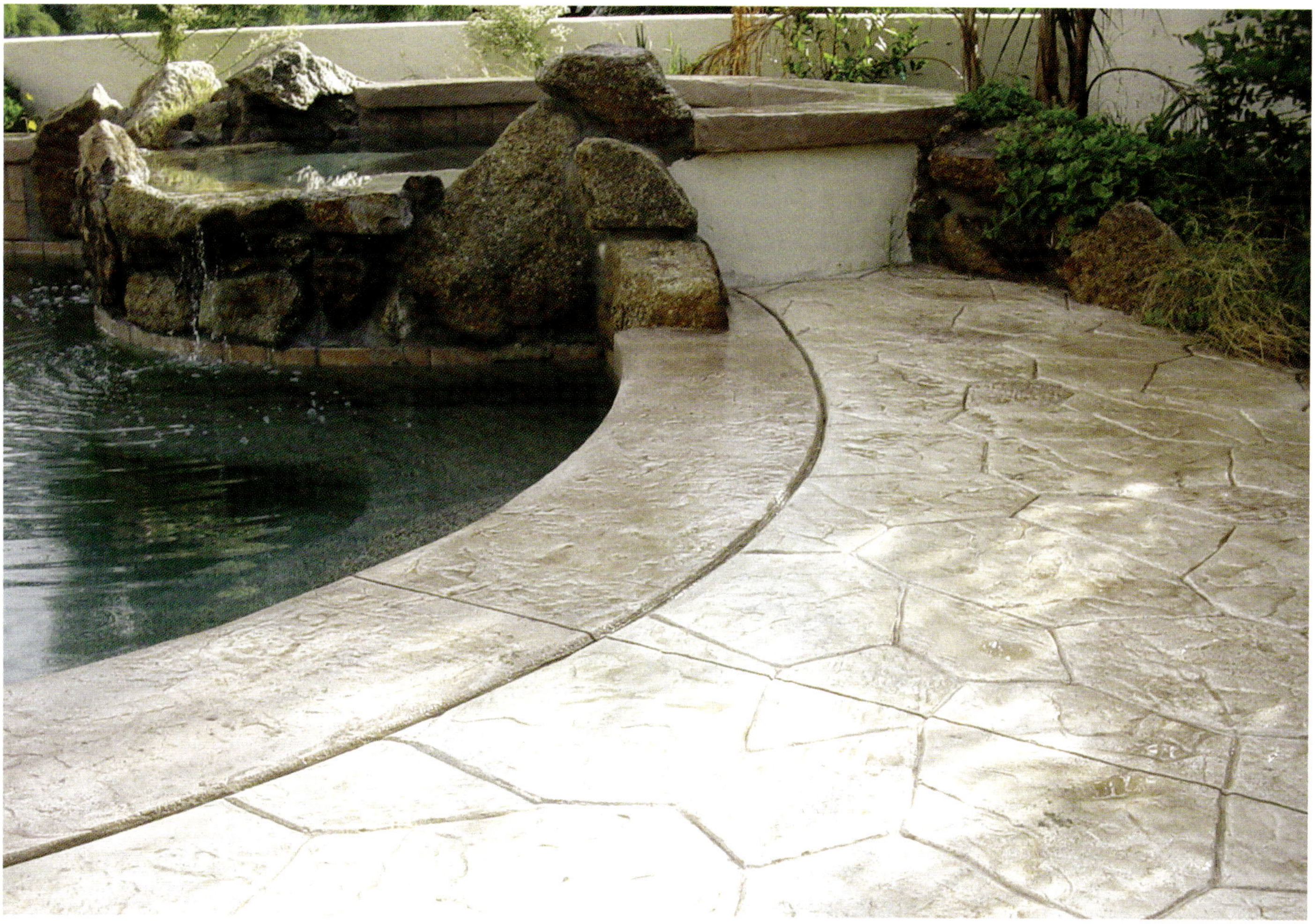

- Use concrete with a low water-cement ratio (see next section, Water-cement ratio: Why lower is better) to make it more durable and less permeable.
- Use a dry-shake color hardener, which will produce a denser, more impermeable surface.
- Tell the owner to avoid using deicing salts, especially during the first winter.

While air in concrete is vital, you need to be careful when specifying an air entrainer for colored concrete. Air-entrainment can significantly reduce bleeding of the concrete (the tendency of some of the water in the mix to rise to the surface of freshly placed concrete). If you're using a dry-shake color hardener, which needs some bleed water to wet out sufficiently so it can be floated into the surface, too much air-entrainment can be a problem. Also be careful when using integral coloring admixtures. Some are formulated to add a small percentage of air (maybe 1 to 3 percent). When they are used in a mix that already contains an air-entraining admixture, you could end up with concrete that's very sticky and difficult to work with. The bottom line: Work with your ready-mix producer to come up with the right balance of air content and workability.

Water-cement ratio: Why lower is better

The quality of concrete is greatly influenced by the water-cement ratio. Through a chemical reaction, called hydration, cement and water form a paste that surrounds and binds the aggregate particles. Adding too much water dilutes and weakens the cement paste, which makes the concrete less durable. Adding too little water makes the fresh concrete difficult to place and finish. The key to producing high-quality concrete is to keep the water-cement ratio as low as possible without sacrificing workability. For exterior concrete exposed to freezing and thawing and deicing chemicals, a maximum water-cement ratio of 0.45 is usually required.

Advantages of keeping water contents as low as possible include:

- Increased concrete compressive strength and durability
- Lower permeability, which improves water tightness and resistance to chemical attack
- Less volume change from wetting and drying
- Reduced shrinkage and cracking

Don't let all your stamping efforts go to waste by using a concrete mix that can't stand up to the conditions it will be exposed to. No sealer or protective coating on the market can compensate for concrete that's not durable to begin with.

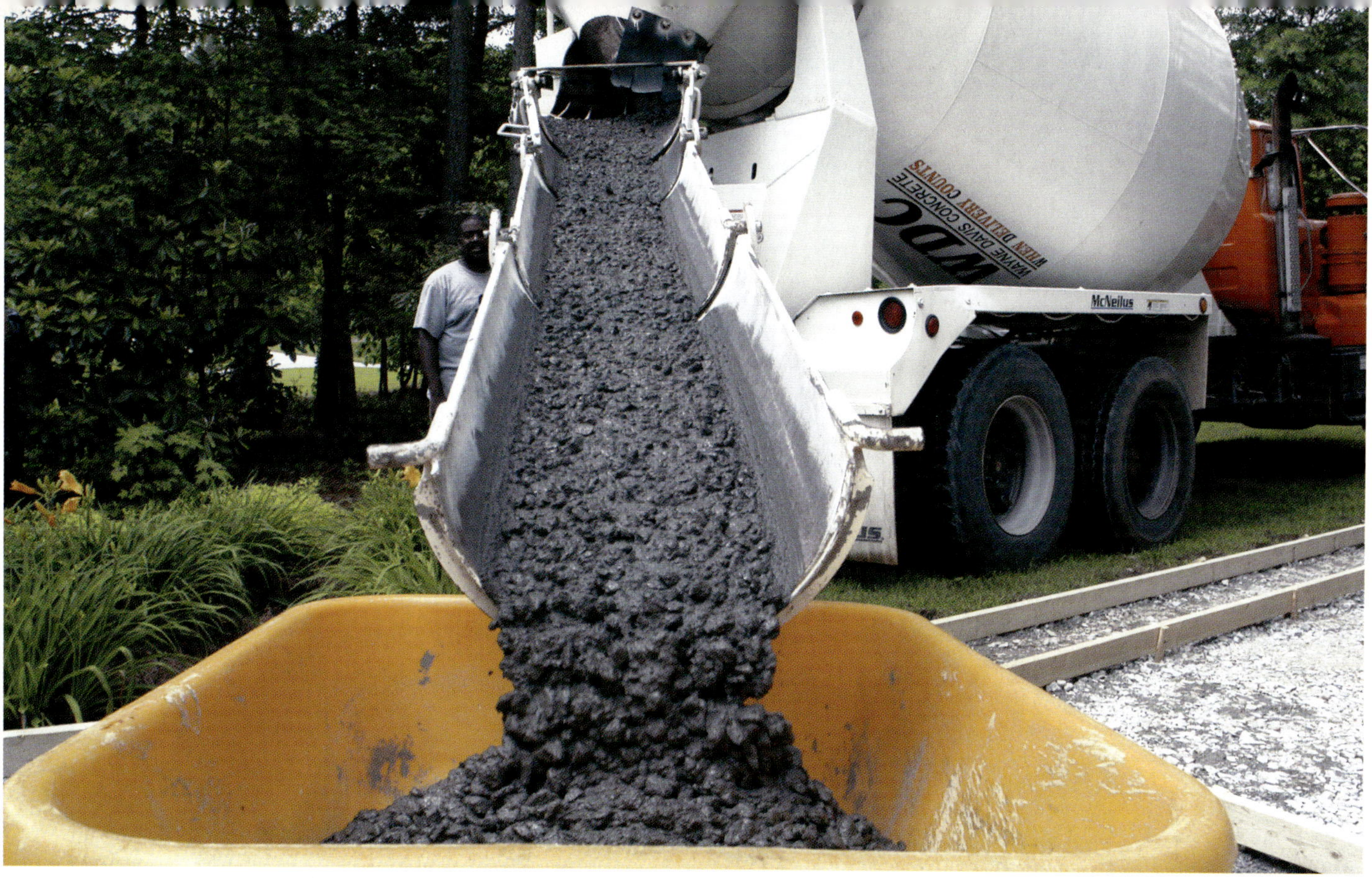

DECORATIVE CONCRETE INSTITUTE

Entrained air in concrete is vital to protect it against freeze-thaw damage. Work with your ready-mix producer to come up with a mix that has the right balance of air content and workability.

TIP

An easy way to calculate the water-cement ratio of fresh concrete is to divide the weight of the mixing water by the weight of the cement. For example, if 1 cubic yard of concrete contains 235 pounds of water and 470 pounds of cement, the mix has a water-cement ratio of 0.50 (235/470 = 0.50). If the mix lists the water in gallons, multiply the number of gallons by 8.33 to determine how many pounds of water the mix contains.

VERLENNICH MASONRY AND CONCRETE

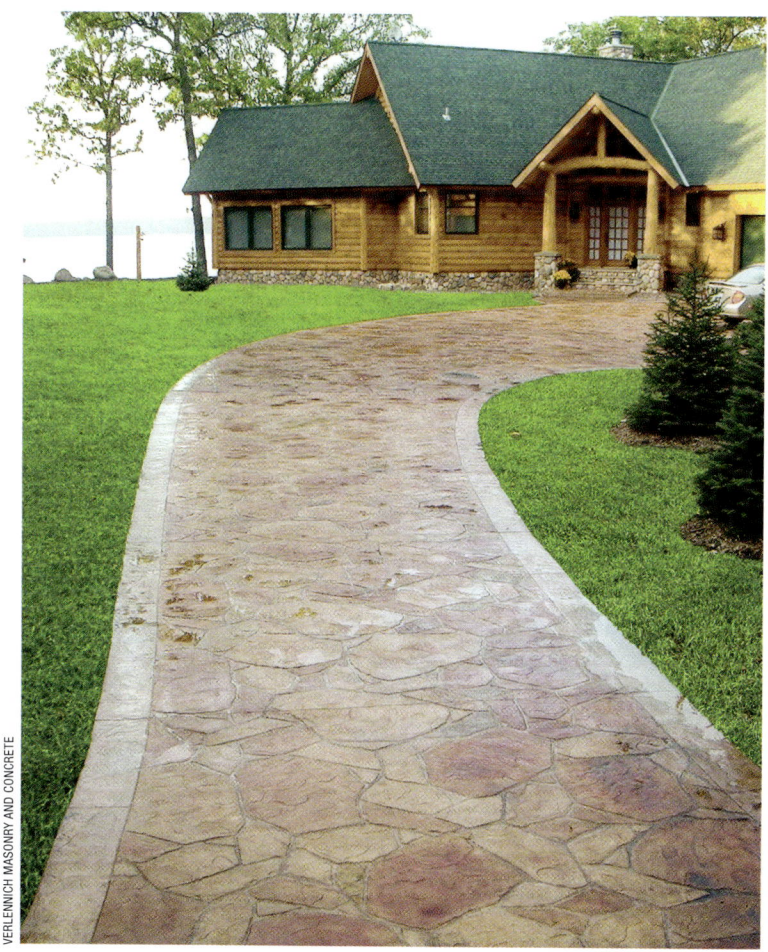

For exterior concrete exposed to freezing and thawing and deicing chemicals, a low water-cement ratio will improve durability and resistance to chemical attack.

CHAPTER 8

ESTABLISHING EXPECTATIONS WITH THE BUILDER, ARCHITECT, AND OWNER

On all decorative stamped concrete projects, whether large or small, you will be dealing with people who have expectations about the final product. On a large commercial job, you often need to satisfy several parties, including the architect, general contractor, and building owner. On a smaller residential job, you'll only have to please the homeowner. In either case, you can build a good relationship and avoid problems by providing a solid foundation of information early on in the project.

Explain the variables

Establishing expectations starts with your initial sales call. This is when you should clearly explain what is—and is not—possible with stamped concrete. Discuss the differences between stamped concrete and other decorative paving materials, such as natural stone. Make buyers aware that stamped concrete is a craft product that's hand-fabricated on site and not a manufactured good, such as tile, that you will have total control over. You are, in fact, dealing with an imperfect material that often must be installed in an imperfect environment, subject to the elements and variable jobsite conditions. That means total continuity and consistency of the final product will be difficult if not impossible to achieve. Variations in the concrete mix design, water-cement ratio, and other factors can also affect the performance and appearance of the final product, as discussed in Chapter 6.

Keep a portfolio of your work

The next step is to show clients the amazing results that are possible with stamped concrete. Present them with a portfolio of your work or refer them to your web site to browse through pictures of previous jobs.

Taking jobsite photos to chronicle your work is of the utmost importance, because each successful installation can become part of the portfolio you use as a sales tool. (This topic is discussed in more detail in Chapter 10, *Chronicling Your Work*.) But if you're just getting started in the stamping business, the brochures of stamping tool manufacturers can be useful until you build up a portfolio of your own. They often are filled with beautiful pictures of projects showcasing various stamp patterns and colors.

Provide samples

Once clients see some of the possibilities, you can work with them to pinpoint a specific pattern and color scheme to be used on their job. At this point, you should explain to customers the differences in the looks attainable with various coloring techniques, such as an integral color versus a dry-shake color hardener. Color charts or color chips can be helpful here, and most manufacturers of coloring products will provide them at no charge for use as a sales tool and to assist in color selection. Keep

CONCRETE IMPRESSIONS

36

Tips for Preparing Samples

Taking the time and effort to prepare representative samples of your work is one of the best ways to ensure client satisfaction. The most important aspect of sample preparation is to make sure your samples accurately represent what you can accomplish in the field.

Two mistakes I've seen contractors make are preparing samples that are too small and failing to show the entire process. I recommend making your samples at least 2x2 feet in size, or large enough to fit two impressions of the stamping tool. This will let clients see how the pattern lines come together and give them a better sense of what the final design will look like. If you're using a texturing skin rather than a stamping mat, you may be able to get by with a smaller sample because no pattern lines are involved.

After stamping the sample, be sure to apply the same color hardener, stains, and acrylic or urethane sealer you propose to use on the job. Eliminating any of these steps or substituting a different product can result in an inaccurate representation of the final installation. Document the formulas used to prepare the sample, and keep accurate notes in a job file so you can replicate what you've done.

Although it's important to prepare samples with care, don't go overboard by using all sorts of elaborate coloring techniques and spending hours to make the sample perfect. That can work against you in the long run because you'll be expected to achieve the same level of perfection over the entire project. It's nearly impossible to be so meticulous and flawless on a concrete placement that's a couple of thousand square feet or larger. Be sure the sample shows what you can realistically achieve in the field.

On large commercial jobs, you may be asked to submit a series of samples or do an actual mockup on the jobsite. Obviously, this can become time consuming and costly. On the other hand, it's much better to find out during the sampling stage whether you are meeting the owner's expectations. I recommend factoring the costs of producing these samples into your bid. Some contractors will provide the first sample free, but charge for any additional samples.

in mind, though, that the color charts are on paper, not on concrete, and most of the color chips manufacturers provide are made from a grout mix in a controlled environment. So use these examples only as a starting point to determine the desired color range.

To help clients narrow down their color and pattern selections, make up actual concrete samples showing some of the combinations they've chosen. Giving clients something they can see and touch, rather than just visualize, will convey the true look and texture of the concrete (see Tips for Preparing Samples on this page).

Also consider opening a showroom or professional design center to showcase the different stamping patterns and color selections you offer. Read more about this topic in Chapter 28, *How to Sell Stamped Concrete Work*.

Get it in writing

Aim for the top in establishing expectations and building that great relationship with your clients—but don't forget to cover your back with the contract. The final decisions you agree on with your clients must be backed up in writing, as discussed in the next chapter.

Display the various stamping patterns and color selections you offer, along with informational flyers detailing the colors and textures used to achieve each look.

One of the best ways to ensure client satisfaction is to show actual samples of your work, prepared using the same stamping tools and coloring techniques you use in the field.

CHAPTER 9

WRITING A FAIR CONTRACT

I have seen so many start-up and even experienced stamped concrete contractors get burned because nothing was put in writing and important issues relating to the project (such as those covered in Chapter 8) were only discussed verbally or not at all. They failed to get a signed contract, one that is fair and covers all the bases.

Such a contract starts with a well-detailed proposal, which you should present to the customer prior to being awarded the project. On a residential job, the client usually signs the proposal and the proposal itself becomes the contract. On commercial projects, most general contractors use their own contract and will expect the stamped concrete contractor to sign it. Make sure this contract incorporates all the important issues included in the proposal you presented to the general contractor.

Here is what a well-detailed proposal addresses:

- Location of the project
- Who the contract is between (the owner, general contractor, owner's agent, etc.)
- A complete description of stamping services to be provided, from site grading to the specified final product
- A list of what is not included
- Provisions for obtaining written approval from the client once a final sample has been agreed upon and before the work starts
- A discussion of the nature of stamped concrete, reinforcing that stamped concrete is not prefabricated and often is not totally uniform or predictable
- Payment schedule
- The need for unrestricted access to the project during grading, forming, and pouring
- Curing times and how long the work needs to be protected
- Start and completion dates of the project, marking the total duration of the stamped concrete work
- Labor rates for standard work hours and overtime or weekend work
- How much advance notice is required prior to starting the project
- Liability responsibilities (general liability insurance, bonding, workers' compensation insurance, etc.)
- Any stipulations for cancellation (for example, a designated time frame for cancellation by either party)
- Site issues such as accessibility, subgrade preparation, drainage, and barricading of the finished work
- A warning stating that even a well executed stamp job, complete with all the necessary isolation and contraction joints, could still develop cracks.

Regarding scheduling, don't set yourself up for failure. Build into the schedule a buffer of a day or two in case work is delayed by rain or other unexpected events beyond your control.

Barring such events, however, if you say you're going to show up to pour concrete on a specific date, be there. Delaying the project without good cause will make you look unprofessional.

Fair contracts are a balancing act: The owner must be given all the important details yet not be scared away by a multipage laundry list of possible problems. Having a fair contract also means being flexible when the unexpected occurs. A true professional will make every effort to work with schedule changes or other obstacles that may crop up.

SAMPLE of A General Construction Contract Form

AGREEMENT: as of the _____ Date: _____
RE: PROJECT ADDRESS: _____
BETWEEN _____
(hereinafter called the "owner") whose mailing address is:

_____

(hereinafter called the "contractor") whose mailing address is:

_____

CONTRACT DOCUMENTS

Contract Documents, which constitute the entire agreement between the Owner and the Contractor and are as fully a part of the Contract as if attached, are enumerated as follows:

(Strike through any that are not applicable to this project).

1. This "Agreement and General Conditions".
2. "Procedures for Contractors".
3. Work Write-Up and Itemized Bid Dated _____ ("Specifications and Bid").
4. General Specification Manual.
5. Addenda No. _____
6. Attached sketches/drawings. _____
7. Owner selection list Dated _____
8. Other _____

THE WORK

The contractor shall perform the entire rehabilitation of the residential structure as described in the contract documents except as indicated as follows to be the responsibility of others:

Scope Responsible Party
_____

TIME OF COMMENCEMENT & SUBSTANTIAL COMPLETION:

The Work shall commence within 7 calendar days of authorization by written Notice to Proceed from the Owner.

The Work shall be substantially completed no later than ___ calendar days from the date of the Notice to Proceed. The Contractor shall be liable for and shall pay the owner $_____ as liquidated damages for each calendar day of delay until the work is substantially completed.

Optional

[If Work is delayed at any time by causes beyond the Contractor's control, then the Contract may be extended for such reasonable time as the Owner's Authorized Representatives may determine.]

OWNER'S REPRESENTATIVE

The Owner's Representative shall be
_____

The Owner's Representative shall:

1. Provide administration of this Contract during construction and throughout the warranty period;
2. Visit the site at intervals appropriate to the stage of construction to determine if the Work is proceeding in accordance with the Contract Documents;
3. Based on evaluation of Contractor's invoices for payment, determine the amounts owing to the Contractor;
4. Have authority to reject Work that does not conform to the Contract Documents;
5. If the Contractor fails to correct defective Work or persistently fails to carry out the Work in accordance with the Contract Documents, by a written order, may order the Contractor to stop the Work, or any portion thereof, until the cause for such order has been eliminated.

CONTRACTOR'S RESPONSIBILITIES

The Contractor shall supervise and direct the Work, using his/her best skill and attention, and he shall be solely responsible for all construction means, methods, techniques, sequences and procedures and for coordinating all portions of the Work under the Contract.

The Contractor warrants to the Owner that all materials and equipment incorporated in the Work will be new unless otherwise specified, and that all Work will be of good quality, free from faults and defects and in conformance with the contract Documents. All Work not conforming to these requirements may be considered defective.

The Contractor shall give all notices and comply with all laws, ordinances, rules, regulations, and lawful orders of any public authority bearing on the performance of the Work, and shall promptly notify the Owner's Representatives if the Drawings and Specifications are at variance therewith.

The Contractor shall be responsible for all safety precautions in connection with this Work. He shall take all legally required and reasonable precautions for the safety of all employees on the Work and other persons who may be affected thereby.

Contractor's liability insurance shall be purchased and maintained by the Contractor to protect him from claims under workers' or workmen's compensation acts and other employee benefit acts, claims for damage because of bodily injury, including death, and from claims for damages, other than to the Work itself, to property which may arise out of or result from the Contractor's operations under this Contract, whether such operations be by himself or by any Subcontractor or anyone directly or indirectly employed by any of them. This insurance shall be written for not less than any limits of liability specified in the Contract Documents, or required by law, whichever is the greater, and shall include contractual liability insurance applicable to the Contractor's obligations under this Section. Certificates of such insurance shall be filed with the Owner prior to the commencement of the Work.

The Contractor shall not employ any Subcontractor to whom the Owner's Representatives or the Owner may have a reasonable objection. The Contractor shall not be required to contract with anyone to whom he has a reasonable objection.

CONTRACTOR "HOLD HARMLESS" WARRANTY

To the fullest extent permitted by law, the Contractor shall indemnify and hold harmless the Owner and the Owner's Representatives and their agents and employees from and against all claims, damages, losses and expense, including but not limited to attorneys' fees arising out of or resulting from the performance of the Work, provided that any such claim, damage, loss or expense (1) is attributable to bodily injury, sickness, disease or death, or to injury to or destruction of tangible property (other than the Work itself) including the loss of use resulting therefrom, and (2) is caused in whole or in part by any negligent act or omission of the Contractor, any Subcontractor, anyone directly or indirectly employed by any of them or anyone for whose acts of any of them may be liable, regardless of whether or not it is caused in part by a party indemnified hereunder.

Such obligation shall not be construed to negate, abridge, or otherwise reduce any other right or obligation of indemnity which would otherwise exist as to any party or person described in this Section. In any and all claims against the Owner or the Owner's Representatives or any of their agents or employees by any employee of the Contractor, any Subcontractor, anyone directly or indirectly employed by any of them or anyone for whose acts of any of them may be liable, the indemnification obligation under this Section shall not be limited in any way by any limitation on the amount or type of damages, compensation or benefits payable by or for the Contractor or any Subcontractor under workers' or workmen's compensation acts, disability benefit acts or other employee benefit acts.

CORRECTION OF WORK

The Contractor shall promptly correct any Work rejected by the Owner's Representatives as defective or as failing to conform to the Contract Documents, whether observed before or after Substantial Completion and whether or not fabricated, installed or completed, and shall correct any Work found to be defective or nonconforming within a period of one year from the Date of Substantial Completion of the Contract or within such longer period of time as may be prescribed by law or by the terms of any applicable special warranty required by the Contract Documents. The provisions of this Article apply to work done by Subcontractors as well as to Work done by direct employees of the Contractor.

CHANGES IN THE WORK

The Owner, without invalidating the Contract, may order Changes in the Work consisting of additions, deletions, or modifications, the Contract Sum and the Contract Time being adjusted accordingly. All such changes in the Work shall be authorized by written Change Order signed by the Owner's Representatives and the Contractor.

CONTRACT SUM/PROGRESS PAYMENTS

The Owner shall pay the Contractor for performance of the Work, subject to additions and deductions by approved Change Orders, the Contract Sum of $_____
_____. The Contract sum is determined as follows:

Base Bid _____

Addenda _____

Contract Sum _____

Based upon invoices submitted to the Owner's Representatives, the Owner shall make payments on account of the Contract Sum to the Contractor as follows:

Draw 1 ____ % $ _____
Draw 2 ____ % $ _____
Draw 3 ____ % $ _____
Draw 4 ____ % $ _____
Draw 5 ____ % $ _____

Payments may be withheld on account of

1. Defective work not remedied,

2. Claims filed,

3. Failure of the Contractor to make payments properly to subcontractors or for labor, materials, or equipment,

4. Damage to the Owner or another contractor, or

5. Persistent failure to carry out the Work in accordance with the Contract Documents.

Final payment shall not be due until the Contractor has delivered to the Owner a complete release of all liens arising out of this Contract or receipts in full covering all labor, materials and equipment for which a lien could be filed, or a bond satisfactory to the Owner indemnifying him against any lien. If any lien remains unsatisfied after all payments are made, the Contractor shall refund to the Owner all moneys the Owner may be compelled to pay in discharging such lien, including all costs and reasonable attorneys' fees. Owner may withhold a retainage of 20% of all invoiced charges if the Contractor fails to complete all contract items upon submission of final invoice.

This Agreement entered into as of the day and year first written above by:

OWNER(S)

_____ (signature)

_____ (signature)

CONTRACTOR:

_____ (authorized signature)

Company Name and Address Goes Here

Items in a standard contract (above) are written from the buyer's perspective and provide protection to the buyer. As the contractor, make sure the items listed on page 40 are included in an addendum that is attached and becomes part of the contract.

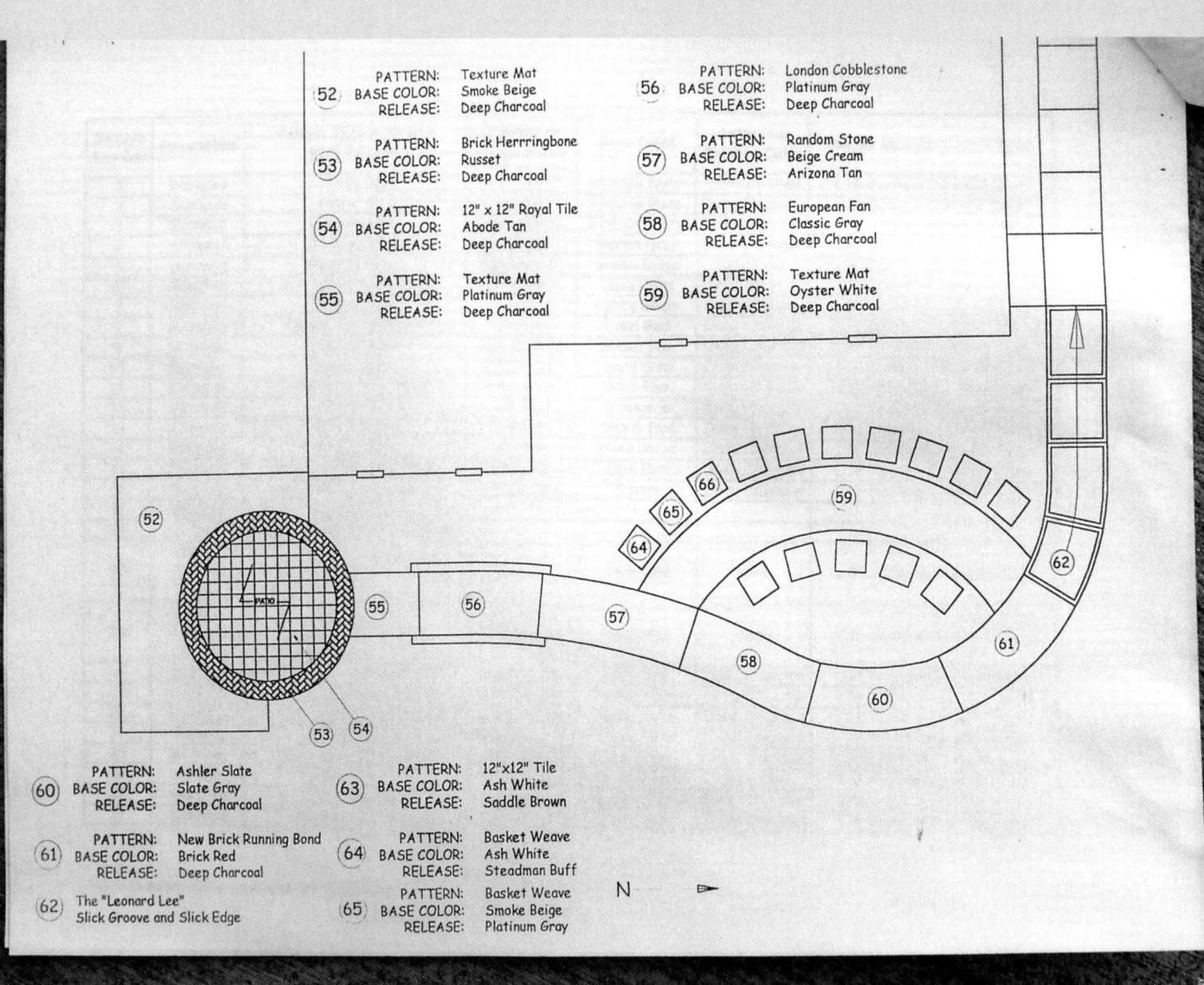

For each job, prepare a detailed legend identifying the colors and stamp patterns used. This "blueprint" is especially useful for documenting complex projects incorporating multiple colors and patterns.

CHAPTER 10

CHRONICLING YOUR WORK

Your completed stamped concrete installations can be your most powerful sales tool. Often potential clients will be wowed by your work and want you to duplicate the same look for their home or business.

But what if the project that caught the client's eye was installed several years ago? Would you be able to recall, off the top of your head, precisely what procedures you followed and the concrete mix design you used?

Probably not. That's why it is essential for you to chronicle your work by keeping detailed records of every project. Not only will this database allow you to retrace exactly what you did on each job, it can also be useful in selling your work and defending against legal claims.

Take notes at every step

Chronicling your work starts with the first set of samples you submit to the client. If you submit multiple samples that have been fabricated with different colors and stamping patterns, label the back of the sample and date it so you can easily identify it later.

Throughout the stamping project, be sure to take accurate notes of the materials you used and in what amounts, including the concrete mix design, dosage of color hardener, accent colors and amount of accenting, and release type and color. If you're using an integral color or a dry-shake color hardener, note the specific color used and the name of the manufacturer. (Different suppliers often use unique color codes.)

If you're blending four or five different dry-shake colors, be very specific in noting the coverage rates used. For example: 100% light-beige base application followed by flash accents of 15% dark brown and 10% terra cotta.

Invite clients to tour your showroom by providing informative flyers describing the many options on display. They can use this information later, during the decision-making process.

Creative Chronicling

L.L Geans Construction Co. of Mishawaka, Indiana, performs decorative concrete work—including stamping and staining—for both residential and commercial clients. Owner Rocky Geans has a showroom filled with samples of his company's work. By the entrance is a mailbox in which he keeps flyers detailing what he did to create each of the samples on display. Potential customers can refer to this "guide" as they tour the showroom to learn exactly what colors and textures were used to achieve each distinctive look.

Don't neglect to take pictures showing the steps before concrete placement, such as subgrade preparation and installation of reinforcement, in case you need to prove later that your crew followed the proper procedures.

Also note the type of sealer used. Was it an acrylic or urethane? Was the finish high-gloss or matte? And how many days after the pour did you apply the sealer?

Keep a camera on hand

It's also a good idea to supplement your notes by taking sequential photographs of the entire project—preferably with a digital camera, so you can easily post the photos on your web site or organize them in a computer database. Many consumers don't understand the stamping process. Showing them photos of each phase—from the pouring of the concrete, to the stamping and coloring process, to final sealing—can help you explain the steps involved. Then you can close the sale by showing spectacular photos of the completed project in a fully landscaped setting. Be sure to keep the best finished shots in your portfolio.

Photos can also protect you from compromising situations or potential lawsuits. Take pictures of the jobsite and any nearby structures before you start your work, especially if other trades are on the project. If a home or building owner later refuses to pay for the work, claiming that your crew ruined their landscaping or damaged their brick or stucco walls, you will have photographic evidence that the damage was already there before you arrived on the project.

Photos may also offer proof that you have complied with local building codes. For example, if the city building code requires the installation of reinforcing steel in concrete pavements, a photo taken before you placed the concrete will show that you did, in fact, install steel in the slab.

Take photos of completed projects in a fully landscaped setting. Keep those exhibiting your best work in a portfolio.

DECORATIVE CONCRETE INSTITUTE

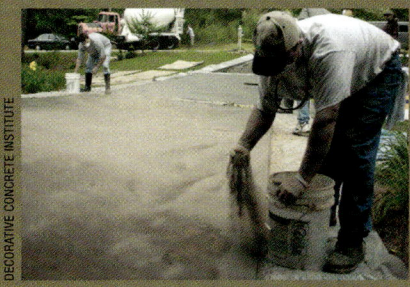

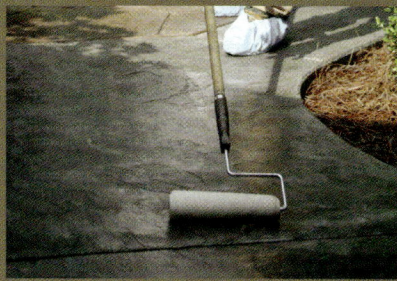

For each stamping project, keep a record of the materials used and in what quantities so you can duplicate the results on other projects. Note the specific colors of dry-shake hardener used and the name of the manufacturer. Also jot down the type of sealer applied (such as an acrylic or urethane) and the finish (matte or high gloss).

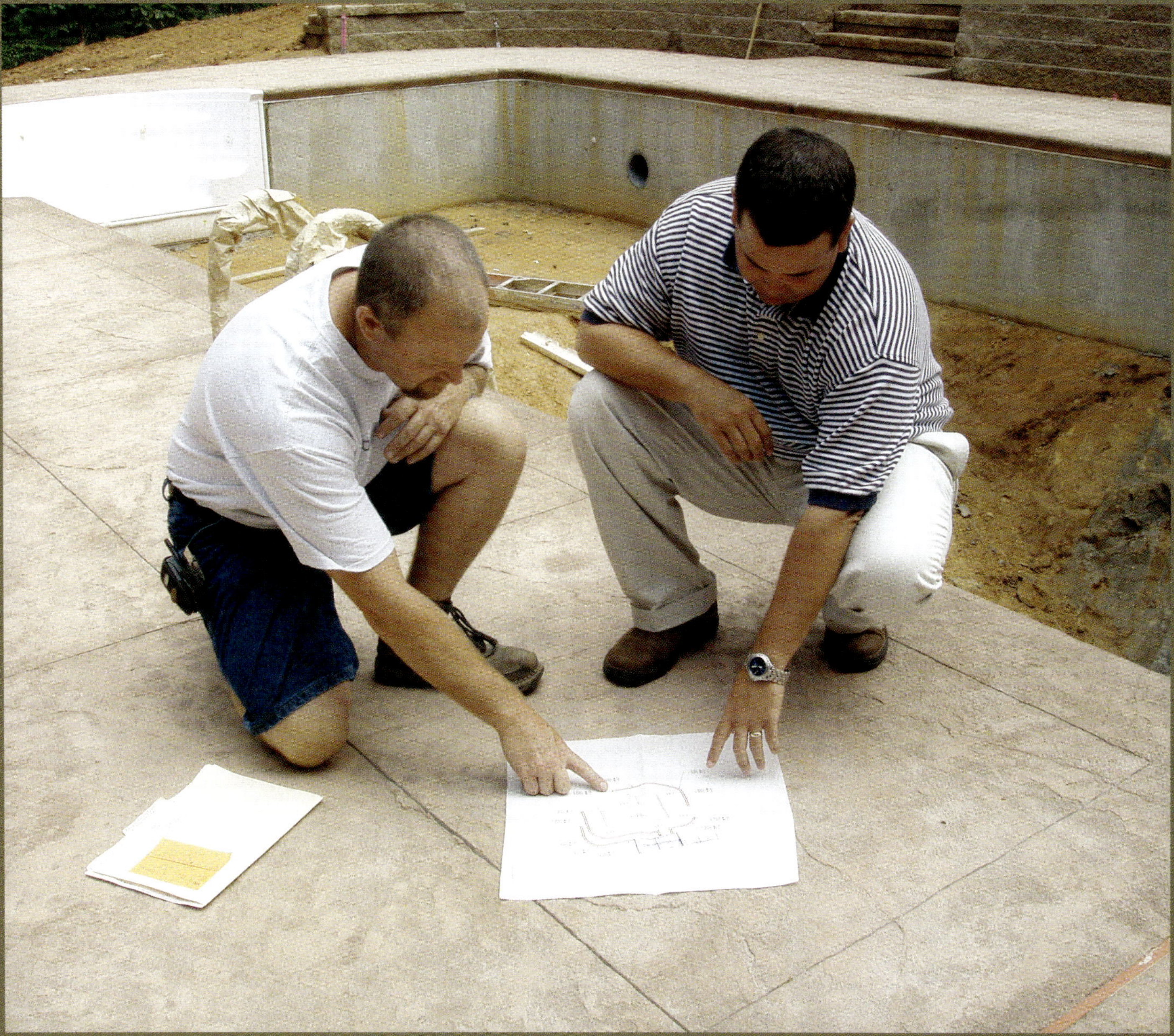

DECORATIVE CONCRETE INSTITUTE

Many of your customers will want to learn more about the decorative stamping process. Explain the steps involved by showing them photos or diagrams illustrating each phase of the work.

CHAPTER 11

THE IMPORTANCE OF SAFETY

It's all too easy to get caught up in the creativity of the stamping process and gloss over safety. Don't do it. Not only can a disregard of safety procedures be harmful to you and your employees, you may also be putting other trades, building occupants, and even the environment at risk. Such negligence can be extremely costly, resulting in hefty fines from the Occupational Safety and Health Administration (OSHA) or the Environmental Protection Agency (EPA).

Safety starts with establishing a comprehensive jobsite safety program and putting someone in charge of overseeing it. This person should keep abreast of all safety rules and regulations, conduct regular safety talks with employees, monitor safety compliance, and make sure the appropriate Material Safety Data Sheets (MSDS) are on site.

There are so many aspects to jobsite safety that I can't possibly cover all of them in detail in this guide. However, here are recommendations for handling potential hazards you may encounter on a stamped concrete project:

- Provide safe access for ready-mix truck drivers, making sure no obstacles are in their way. If you are signaling a driver to back up to deposit the concrete, make certain he can see you in the rearview mirror. Never stand behind the concrete truck or squat down. Also make sure other workers are clear of the area.

- Once the truck begins unloading the concrete, don't walk underneath the fully loaded delivery chute. On rare occasions, I've seen the hydraulics break and the chute come slamming down to the ground.

Be careful when standing near a fully loaded delivery chute. Never turn your back to it or walk beneath it.

- Take precautions when working with fresh concrete. The cement in concrete is very caustic and can cause severe skin burns, even after brief contact. Always protect your hands with nonabsorbent gloves and wear tall rubber boots to protect your feet, ankles, and legs from wet cement. Also wear safety glasses to keep concrete splatter out of your eyes.

- Protect your knees from wear and tear by wearing kneepads whenever you are on the slab finishing the concrete.

- Bags or buckets of color hardener are heavy, sometimes exceeding 60 pounds. Learn to lift properly (bending at the knees and lifting with your legs) so you don't throw out your back.

- When you work with dry-shake color hardeners and powdered release agents, some of the material goes airborne and can be very harmful to breathe. Be sure to wear a respirator or dust mask, per OSHA regulations, to protect your lungs.

- If you are finishing the concrete using a bull float or fresno with a long aluminum handle, be careful of any overhead electrical wires. You could be electrocuted if the metal contacts exposed wires. A good way to avoid this problem is to use a fiberglass handle.

- Solvent-based acrylic sealers are extremely flammable and can be hazardous to breathe. Keep all possible sources of ignition away from the work area, such as lit cigarettes, open flames from a space heater, or sparks from a power saw or other equipment. If you're working indoors, make sure the area is well ventilated.

- When using power equipment, always follow the recommended safety procedures for operation and wear the proper safety gear, such as safety glasses, work gloves, and hearing protection.

- Check extension cords to make sure they are properly grounded and have no cut or frayed portions. Never use electrical equipment around water.

- Use caution tape or some type of barricade around the work area to reduce the risk of trip-and-fall accidents. Formwork and exposed rebar can be particularly hazardous.

- Don't ignore environmental safety. Capture and dispose of any hazardous residue left behind from your work. This is getting to be a very serious problem in some parts of the country. Protecting the environment should be a crucial aspect of any jobsite safety program.

When kneeling on the slab to finish the concrete, wear kneepads for greater comfort and to prevent strain.

When working with fresh concrete, always protect your hands with nonabsorbent gloves and wear tall rubber boots to protect your feet, ankles, and legs.

Safety Resources on the Web

American Society of Concrete Contractors safety mentors page:
www.ascconline.org/Newsite/safety_mentors.htm

American Society of Safety Engineers: www.asse.org

Environmental Protection Agency: www.epa.gov

National Institute for Occupational Safety and Health: www.cdc.gov/niosh

Occupational Safety and Health Administration: www.osha.gov

SafetyInfo (a library of online safety resources): www.safetyinfo.com

Save your back. Don't overload wheelbarrows with more weight than you can comfortably handle.

Dry-shake color hardeners and powdered release agents can be harmful to breathe. Always wear a respirator or dust mask to protect your lungs from any material that drifts your way.

CHAPTER 12

STAMPED CONCRETE PICTORIAL OVERVIEW

Why are more and more homeowners, businesses, and municipalities choosing stamped concrete to enhance their landscapes, buildings, and communities? The answer is evident in this collection of photos. No other paving material matches the versatility of stamped concrete by offering such a wide array of patterns, textures, and colors.

Innovation and quality workmanship also play important roles in the surging popularity of stamped concrete. The advances in stamping tools and coloring products over the past decade have made it possible for skilled concrete contractors to simulate virtually any paving material—including natural stone, brick, glazed tile, and slate—and usually at a much lower cost than the real thing.

Special thanks to the product suppliers and talented contractors who contributed the photos shown here and elsewhere throughout this guide. They provide striking examples of the enduring beauty of stamped concrete

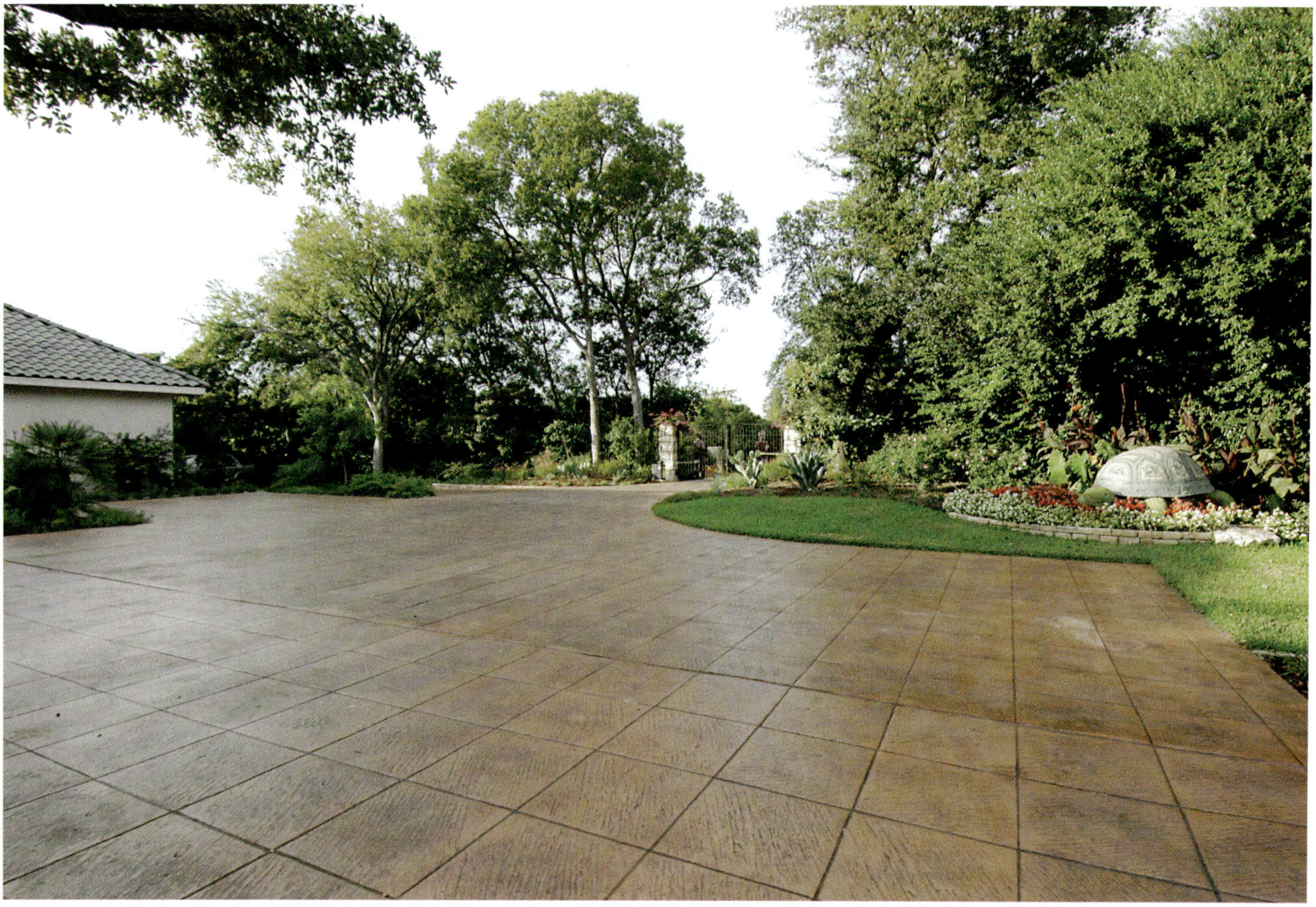

CONCRETE IMPRESSIONS

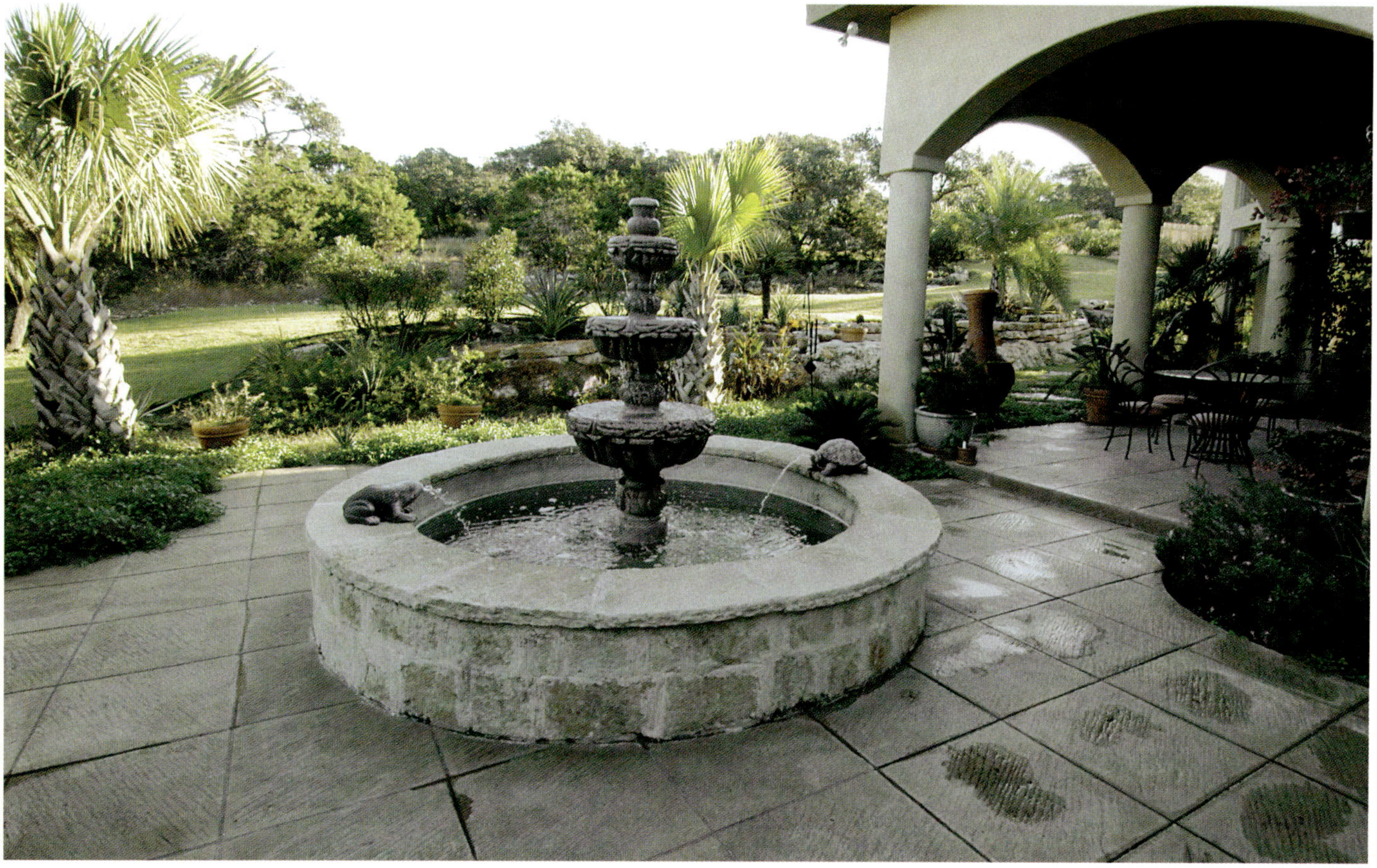

CONCRETE IMPRESSIONS

CHAPTER 13

METHODS OF COLORING STAMPED CONCRETE

Stamping mats and skins impart texture and pattern to concrete, giving it the look and feel of stone, brick, slate, and other rough-textured materials. But to complete the effect, you also need to replicate the natural colors of those materials.

Stamping contractors have an array of coloring methods to choose from. Typically, they will start with a base color produced by either an integral color or a dry-shake hardener. Then to achieve a more natural, variegated appearance, they often apply accent or antiquing colors using pigmented powdered or liquid release agents, possibly supplemented by stains, dyes, or tints. Experienced contractors often combine several of these coloring methods to achieve the most realistic effects. Here is an overview of the various coloring options available and some of the advantages and disadvantages of each.

Integral color

Integral color is easy to use because it is typically mixed right into the concrete at the batch plant to achieve uniform, homogeneous color. Integral colors are available in powdered, granular, or liquid forms, and sometimes contain admixtures, which is why they are also called integral coloring admixtures. Some concrete batch plants have liquid dispensing systems that use computer-controlled machinery to meter and dispense the liquid pigments to improve dosage control and color consistency. Other batch plants have pre-packaged, pre-measured pails of liquid color dosed in cubic-yard increments, which are then dispensed into batches of concrete in the ready-mix truck.

A big advantage of using integral color is labor savings. You don't need to float the color into the surface during finishing, as you do with shake-on hardeners. Another plus is that the color is permanent because

Integral coloring admixtures are usually mixed into the concrete at the batch plant. To improve dosage control and color consistency, some plants use computer-controlled machinery to meter and dispense the pigments.

Shake-on hardeners are broadcast onto the fresh concrete and then floated into the surface before imprinting. They are a blend of color pigments, portland cement, finely graded silica sand, and wetting agents.

With dry-shake hardeners, you can attain rich, vibrant color in an unlimited range of hues.

it extends throughout the entire matrix of the concrete. So even if the slab surface is accidentally chipped, scratched, or abraded, the integral color will remain, unlike with surface-applied treatments. Most of today's integral coloring admixtures use chemically stable synthetic mineral-oxide pigments that stand up to exposure to ultraviolet light without fading or changing hue.

A disadvantage with integral color is that the hues are more subtle and less vibrant than what you can achieve with color hardeners. You're generally limited to soft earth tones, such as muted browns, reds, and tans. Although you can obtain pastel hues with integral color, such as blues or greens, doing so is usually cost prohibitive. To achieve good color intensity with these lighter hues, you would need to use a white cement and a high dosage of pigment, resulting in a significant increase in concrete costs. For that reason, stamping contractors often use integral colors in conjunction with surface-applied treatments—such as color hardeners and chemical stains—to create layers of color.

Dry-shake color hardeners

For brighter colors, dry-shake color hardeners give better results and come in an unlimited range of color options.

Most shake-on hardeners are a blend of color pigments, portland cement, finely graded silica sand, and wetting agents. The hardeners are broadcast onto the fresh concrete and then floated into the surface before imprinting. To work properly, the color hardener must "wet out," or absorb some moisture from the slab (see Chapter 19, *Applying Color Hardener*). Because these products contain fine aggregate and cement, as well as other ingredients, they actually densify the surface and make it less permeable, so some surface strengthening can be expected.

While dry-shake hardeners are more labor-intensive to use than integral coloring, they are comparable in price overall with integral color because you're not coloring the whole matrix of the concrete. Hardener is applied only in the quantity needed to color the top 1/8 to 3/16 inch of the slab.

Powdered or liquid release agents

Pigmented powdered or liquid release agents serve dual purposes: They act as bond breakers to prevent the stamping mats or skins from sticking to the concrete and disturbing the imprint texture, and they impart subtle color to the concrete that enhances the integral or dry-shake

> **Tip**
>
> Generally, 1 cup of powdered release is needed to tint approximately 5 gallons of liquid release. A helpful hint: Add the powder in increments, maybe 1/2 cup at a time, until the tint level you want is achieved. Mix the powder with the liquid release, and then dip a clean rag, sponge, or brush into the tinted liquid and drag it across a piece of white paper to check the color intensity of the tint. If you decide you need to intensify the tone, add another small dose of powder. It's much easier (and less wasteful) to build up the tint level this way than to add more liquid release to soften a tone that's too bold.

color, resulting in an antiquing effect.

A popular technique is to start with a light concrete base color (whether an integral color or a color hardener) and then apply a much darker release agent for contrast. Although about 70 to 80 percent of the powdered release is washed away after the concrete hardens, the remaining release becomes depressed into the surface paste during stamping, which creates the subtle color accents.

Powdered release agents are more traditionally used by stamping contractors because they offer more color selections. However, they do have disadvantages. Because these very fine powders are dusted onto the concrete surface, they create airborne dust particulates, so workers must wear dust masks to prevent inhalation. On windy days, the airborne powder can stain nearby buildings, existing concrete flatwork, and landscaping, making it necessary to mask off adjacent areas with plastic or paper sheeting. Because of these concerns with powdered releases, more contractors are using clear liquid releases and tinting them with a powdered release (see the tip on page 59). This method is quite effective and eliminates the disadvantages of using a powder alone.

I'm from the old school, though, and believe you can obtain a much more realistic look using a straight powder release. It's often a tradeoff: Powders require more cleanup and masking of adjacent surfaces, but they produce greater contrast than a liquid release.

Stains

Some stamping contractors like to achieve color variegation by applying chemical stains to the concrete after it has cured. Acid-based stains react chemically with the concrete and produce a mottling effect that gives your stamping work a sense of realism. For example, if you're trying to mimic the color variations of natural stone, an acid stain will allow you to achieve interesting highs and lows. Most stains can be diluted to achieve varying degrees of color transparency. You can also apply stain randomly to individual stones in the stamped design—something that would be impossible to do with color hardeners or releases that are worked into the concrete surface.

Another reason to use a stain is to mask blemishes. No matter how careful you are, it is difficult to produce a blemish free stamp job. There will always be some subtle color imperfections. For example, maybe the pigmented release didn't take completely in one area. You can use a stain to accent the color in that spot and disguise the flaw.

Dyes and tints

Dyes can be used in conjunction with stains to achieve greater color intensity.

You can apply chemical stains randomly to individual stones in a stamped design to achieve realistic color variations. The stain is applied after the concrete has cured and can be layered over integral or dry-shake color.

They can produce bright, vibrant colors, and you can mix your own custom colors on the jobsite to broaden your color palette.

Unlike stains, dyes are not chemically reactive with concrete. Instead, the fine coloring agents in dyes penetrate the concrete surface. Keep in mind, however, that dyes will fade somewhat (generally 5 to 10 percent) when exposed to ultraviolet light. They are not as UV stable as acid stains.

Tints, which are generally diluted color washes, can be used to add hints of color and produce some interesting faux finishing effects. There are several methods of creating tints. Probably the most popular is to mix a pigmented powdered release with a solvent-based acrylic sealer, so your sealer itself acts as a color wash. A word of caution: This method won't work with water-based products; you need the solvent to break down the particles in the powdered release.

I've also seen contractors make tints by mixing several handfuls of color hardener with water in a pail. The heavier solids in the hardener settle to the bottom of the pail while the pigments color the water. The tinted water can then be applied to the concrete by spray or sponge. Keep in mind, however, this type of tint is not as permanent as a stain or color hardener. You will definitely need to lock in the color by applying several coats of sealer.

A popular technique for antiquing surfaces is to start with a light concrete base color and then apply a darker release agent for contrast.

Tints are diluted color washes that you can mix yourself to add just a hint of color.

A well-prepared subgrade is firmly compacted to prevent settlement of the stamped concrete that will rest on top of it.

For compacting subgrades on smaller residential jobs, a vibratory plate compactor works well, supplemented by a hand tamper for use in tight areas.

Before placing the concrete, check that the subgrade is level and uniformly graded.

CHAPTER 14

SUBGRADE PREPARATION

One of the most critical aspects of good concrete construction is also one of the most overlooked: proper subgrade preparation. As discussed in Chapter 5, the subgrade can make a big difference in the overall performance and structural integrity of the concrete slab. A well-compacted subgrade also helps drainage and can prevent soil erosion under the concrete.

On commercial projects, an engineer may provide a soils report along with specifications detailing the requirements for subgrade preparation, such as the amount and type of aggregate fill to use as the base. Follow these guidelines closely.

In noncommercial settings or in cases where no soils report or specifications are provided, you may be asked to provide your own recommendations. Spell out in your contract what your intentions are for preparing the subgrade based on past successful installations. I know of a contractor in California who got into trouble when he installed an 8,000 square foot driveway that later heaved due to unstable soil. No soils report had been provided to him, but the owner held him liable anyway.

If you are asked to install the subgrade per your recommendations, here are some general guidelines to follow that have worked well for me:

- Match the thickness of the crushed, granular fill to the thickness of the concrete slab. For example, if you're pouring a 4 inch thick driveway or patio, use a minimum of 4 inches of compacted aggregate. The same holds true for thicker load-bearing slabs that must support heavy vehicular traffic, such as bus turnarounds and crosswalks. For these slabs, the concrete may be as thick as 8 to 10 inches. Put down the same thickness of compacted aggregate fill.

- All fill must be well compacted to prevent settlement. If the subgrade settles, it will leave the concrete slab only partially supported. A roller compactor works well on large projects. For smaller areas, I like to use a vibratory plate compactor. A hand tamper also works well, but is extremely labor intensive.

- To ensure proper consolidation, install the aggregate fill in lifts. For instance, if you're putting down 10 inches of material, moisten and compact 2 to 4 inches at a time. Moistening will facilitate compaction.

- Before placing the concrete, check that the subgrade has no standing water on it or spots that are muddy, frozen, or soft and spongy.

- Make sure the subgrade provides for a uniform thickness of concrete. If the subgrade is not uniformly graded (for example, placed at only a 3 inch thickness in one area and a 5 inch thickness in another), the concrete will not set at the same rate and you could get plastic shrinkage cracking during the stamping phase.

Special Precautions

In coastal regions, contractors will sometimes pour concrete slabs over the top of a sand subbase.

However, it can be difficult to keep sand flat and level during concrete placement, and if the sand shifts as the concrete is placed, you won't get a uniform slab thickness.

The American Concrete Institute (ACI 302.1R-96, "Guide for Concrete Floor and Slab Construction") recommends using a compactible, easy-to-trim granular fill that will remain stable and support construction traffic. There are many different types of granular fill materials, ranging from crushed limestone to granite. I suggest that you do some research and find a suitable material that's readily available in your area.

Whatever fill material you use, make sure it's properly compacted. The subgrade should be able to support a fully loaded concrete truck without the truck wheels making ruts more than 1/4 inch deep. If the truck leaves deeper ruts in the subgrade, they will end up getting filled with the fresh concrete. As the concrete sets, it's restrained in those areas, which can promote cracking.

CHAPTER 15

ERECTING THE FORMS

The type of form most commonly used for concrete flatwork is a standard wood 2x4. Reusable plastic forms are also available. If the slab has curved sections, you can use a thinner 1x4 to form the curve. For tight radii, Masonite forms, made from pressed wood fibers, are very effective. Occasionally, you will be asked to include curbs, which may require forms as large as 2x12 inches or plywood.

The procedures you'll use to erect forms for stamped concrete won't differ much from those used for conventional concrete slabs. But with stamped concrete, it's particularly important not to take any shortcuts. A sloppy forming job can make stamping more difficult and even prevent the stamp pattern from lining up properly.

To prevent headaches and ensure a beautiful stamping job, don't overlook the following basics when erecting forms.

Drive the stakes flush

Be sure to drive or cut off the form stakes flush with, or slightly below, the top edge of the forms. When stamping, time is of the essence. If the stakes are higher than the top of the forms, they will interfere with the stamping mats when you near the edge. The use of flexible mats and hand chisels would then be needed to complete the pattern.

Cut off the form stakes flush with or slightly below the top edge of the forms so they won't interfere with the stamping mats when you near the edge.

Measure elevation carefully

Set the forms to the correct elevation of the slab. That means taking precise measurements using a transit or laser level. For smaller slab sections, you can use a 4 foot carpenter's level along with a straightedge. The proper slope also must be established to ensure adequate water runoff. The amount of slope needed depends on the depth of the stamp pattern and site conditions (as discussed in Chapter 5).

To determine elevation and slope, set grade stakes at the perimeters of all the form boards. Then, using your level, establish the grade and mark the elevations on the grade stakes with a pencil. Tie a string line from stake to stake at the marked elevations to use as a guide for setting the forms. If the slab abuts an existing structure, such as a building or wall, you can mark the elevation on the structure itself using a chalk line.

Be aware that most form lumber has a natural bow or bend to it. A good routine is to sight down the edge of the board and place the bow or bend away from the string line so it does not crowd the line, which could throw the alignment or elevation off. Forming in this fashion allows you to pull the board to the string line while simultaneously driving the stake into the ground. Also, most lumber has a natural crown to it. I prefer the crown up as opposed to down when setting forms. Should the form be slightly higher (relative to the

A stairway becomes a grand entrance when stamped to match adjacent flatwork. Careful attention to form erection and removal facilitates the texturing and coloring of stair treads and risers.

string line), you can then tap it down after the stake has been attached to the form. This will help strengthen and secure the formwork.

Square the forms

For jobs using repeating square or rectangular stamp patterns, such as bricks or tile, the forms must be perfectly squared at the corners. Otherwise, you'll end up with partial bricks or tile running along the form edge, with the pattern looking progressively worse as you move down the slab and away from the corner.

You can check squareness by measuring diagonals of rectangular sections (they should be equal) or by using a 3-4-5 triangle at the 90 degree intersection of the forms. With the latter method, mark one form 3 feet from the end of the corner and the intersecting form 4 feet from the corner. Then measure diagonally between the two marks. The resulting measurement should be 5 feet if the 90-degree intersection is square. This formula can be used in larger increments on larger areas (for example 6-8-10 or 12-16-20). On smaller areas, you can simply use a framing square.

Install adequate bracing

In addition to using upright stakes to hold the forms at the proper elevation, you must install diagonal braces to keep the forms in alignment when the weight of the fresh concrete pushes against them. Wood and steel stakes are the most widely used.

Steel stakes are usually the best choice if you are working in firm soil. The steel won't fracture or split, as wood can, when you pound the stakes into hard ground. The steel will also remain rigid without bending.

Protect the forms after erection

Because it's so important for formwork to be set to the proper elevation and securely braced, you don't want other trades on the project to disturb the forms after they have been erected. This can be a big problem on commercial projects, when other trades are performing their work while you are trying to execute your job. I have seen carefully erected forms destroyed by workers accidentally driving over them or bumping them with heavy equipment. Unless you have sole access to the project, barricade the work area to avoid the time and cost of repairing damaged forms.

Use a level to confirm that forms are set to the correct elevation of the slab.

Check the squareness of forms at corners, especially if you plan to use square or rectangular stamp patterns.

Upright stakes hold the forms at the proper elevation while diagonal braces keep the forms in alignment during concrete placement.

Special Forming Challenges

STEPS

Step risers can be tricky to form if you intend to color and texture the face of the step while the concrete is still plastic. If you strip the face forms too soon, the concrete can sag. If you strip them too late, you won't be able to obtain sufficient color and texture.

The form face should be erected in a fashion that permits easy removal without damaging the plastic concrete. I prefer to use duplex nails rather than screws to attach the forms because they are easier to remove. After removing the form stakes and nails, gently tap the form face straight down with a hammer to break the bond with the concrete. Then gently pull the form upward rather than outward. If you pull the form straight out, you could end up removing a big chunk of concrete. A light mist of water applied to the form before the concrete is placed will also help to break the bond with the concrete. Avoid the use of form release oils, which can discolor the concrete and act as a bond breaker if a dry-shake color hardener is used.

Another option is to use premolded, textured form liners, attaching them to the inside faces of the step riser forms before the concrete is poured. This method generally works best with integrally colored concrete.

To make it easier to float and stamp the stair treads, some contractors put a 45 degree angle cut in the bottom of the face board.

MULTILEVEL DECKS

On multilevel decks where the stamp pattern must align, you need to set benchmarks to make sure the stamp pattern is staying on track and doesn't shift or move. On projects such as this, some contractors mark or lay out the pattern using a grid of freestanding boards. For example, if your stamp mats measure 2x2 feet, lay out a grid with these dimensions to check alignment as stamping progresses.

This is more difficult to do with some stamp patterns, such as a European fan design, especially if you're stamping around a pool deck. The pattern often won't align perfectly once you reach the point where you started. The use of texturing skins and some hand chiseling may be needed to tie in this intersection. Or consider using prepoured concrete borders as start and stop points, allowing you to turn the stamp pattern.

Premolded form liners can be used instead of stamps to impart texture to step faces. Here, a liner with a cantilevered edge is attached to the inside face of a step riser form.

If you plan to color and texture step risers, erect the forms so that the faces are easy to strip while the concrete is still plastic enough to take an impression. Removal will be easier if you use duplex nails rather than screws to attach form faces.

CHAPTER 16

INSTALLING REINFORCEMENT

While decorative stamped concrete is highly prized for its aesthetic appeal, it often serves an important structural function as well, especially if the slab will support vehicle traffic. Reinforcing steel, in the form of reinforcing bars or welded wire mesh, can boost the structural performance of slabs on grade and provide other important advantages (see Reasons to Use Reinforcing Steel on page 71).

A disadvantage of steel reinforcement is that it doesn't begin to offer benefits until the concrete hardens. That's why some contractors also add synthetic fibers to decorative stamped concrete to provide secondary reinforcement. These tiny, hairlike fibers add little or no strength to the concrete, but they do help to control early cracking while the concrete is still plastic. They also contribute some benefits after the concrete hardens (see Reasons to Use Synthetic Fibers on page 69).

On most commercial and public-works projects, reinforcing requirements will be dictated in the specifications provided by the architect or engineer. On residential jobs, the homeowner will typically depend on your expertise in choosing the most appropriate type of reinforcement. Before making a decision however, check with local building code requirements.

Here's a brief overview of reinforcement options.

Steel rebar

Steel reinforcing bars (or rebar) are an effective and economical method of concrete reinforcement. Rebar serves several functions. While concrete has a high compressive strength, it has a very limited tensile strength. Steel bars can overcome these tensile limitations. They can also greatly reduce the amount of cracking in hardened concrete. If you do get a crack and the rebar is properly spaced and positioned, it will often stop the crack from opening any wider.

For concrete flatwork, the typical steel arrangement is #3 or #4 bars spaced 18 or 24 inches on center each way. In most cases, I use #4 bars at an 18 inch spacing.

Welded wire mesh

Like steel rebar, welded wire reinforcement is primarily used in slabs to control cracking caused by shrinkage, thermal stresses, and other effects. However, I'm not a big fan of using welded wire because it's much harder to keep it at the proper depth in the slab (see the tip on this page). Even when the mesh is supported, it usually isn't stiff enough to take the weight of workers who are walking in the fresh concrete to place and finish it.

> **TIP**
>
> To be effective, reinforcing bars and wire mesh must be positioned at the proper depth in the slab. Generally, reinforcement should go in the middle of the slab to one-third the depth from the top. Under no circumstances should it go below mid-depth. To keep the steel in position, be sure to use support devices that can keep the bars or mesh in place during the construction process.

For concrete flatwork, the typical steel arrangement is #3 or #4 bars spaced 18 or 24 inches on center each way. Use tie wire to secure the steel together where it intersects.

Sometimes the reinforcement ends up on the bottom of the slab or touching the subgrade, where it does little good.

Synthetic fibers

When you use synthetic fibers in conjunction with steel reinforcement in concrete you can greatly reduce the amount of cracking that would normally occur. The steel provides the tensile muscle, and the fibers help prevent early shrinkage cracking while the concrete is still plastic. If plastic cracking does occur, the fibers can limit the crack to a hairline or thinner width. However, they won't limit crack width in hardened concrete. That function is provided by the primary reinforcing steel.

If specification or code requirements call for the use of fibers, I usually use polypropylene (although nylon and polyester fibers are also available). These synthetic materials are noncorrosive and can withstand the alkaline nature of concrete. The fibers are added to the concrete at the batch plant to ensure uniform distribution throughout the mix.

Synthetic fibers are very fine, so they are virtually invisible in the slab and won't detract from the appearance of the stamped concrete. But they sometimes present a problem when I'm broadcasting dry-shake color hardener on the surface (see Chapter 19, *Applying Color Hardener*). You need a bit of bleed water to wet out the color hardener, yet the fibers can reduce bleeding, depending on the amount used. Sometimes you can overcome this problem by slightly reducing the dosage of fibers used per cubic yard of concrete. However, be sure to discuss this option with the fiber manufacturer to ensure that the fibers will still provide the desired benefits at the lower dosage.

Reasons to Use Synthetic Fibers

- To reduce plastic shrinkage cracking.
- To reduce plastic settlement cracks.
- To improve the impact, abrasion, and shatter resistance of hardened concrete.
- To reduce concrete permeability.
- To ensure uniform distribution of secondary reinforcement.

Reference: *What, Why, & How: Synthetic Fibers for Concrete (Concrete In Practice No. 24)*, published by the National Ready Mixed Concrete Association (www.nrmca.org).

Reinforcing steel gives stamped concrete driveways (and other slabs exposed to vehicle traffic) the tensile strength needed to support heavy loads. Another benefit: The slab will be less likely to crack.

Use supports to keep the steel bars or wire mesh in place during concrete placement. The reinforcement must be embedded mid-depth or slightly higher in the slab to be effective.

Reasons to Use Reinforcing Steel

- To control cracking caused by drying shrinkage, temperature changes, and applied loads. When properly installed, the steel prevents cracks from opening wider and becoming objectionable.
- To permit wider joint spacings (depending on the design and intended purpose of the slab).
- To bridge any soft spots in the subgrade and provide added structural capacity.
- To increase impact resistance.
- To reduce slab and joint maintenance by keeping cracks tightly closed.

Reference: *Reinforcing Steel in Slabs on Grade* (Engineering Data Report No. 37), published jointly by the Concrete Reinforcing Steel Institute (www.crsi.org) and Wire Reinforcement Institute (www.wirereinforcementinstitute.org).

CHAPTER 17

PLACING THE CONCRETE

There are a number of ways you can place the concrete once it arrives at the worksite. The method you choose is often dictated by site restrictions, size of the job, labor resources, and budget (see Getting Concrete Where You Want It on page 74). Be sure to determine the best method of concrete placement before the job begins so you have the necessary equipment and labor on hand.

Keep the pathway for the concrete truck free of obstructions so the driver can back up to the placement area without interference.

Unless site access is a problem, the easiest approach is to have the ready-mix truck simply pull up to the placement area and deposit the concrete right from the chute (called "tailgating"). Map out a pathway for the truck and keep it clear of obstructions so the driver can pull into the placement area without interference.

If you plan to use a small-line pump to place the concrete, be sure to tell your concrete supplier so he can match the mix to the pump's capabilities. Among the factors to consider are the diameter and length of the pump line. Also, have enough workers on hand during the placement to take advantage of the pump's faster delivery rates.

If the concrete will be placed by a motorized buggy or wheelbarrow, and you have reinforcing steel in the slab, you will need to assemble a ramp that will allow the buggy to ride over the top of the steel.

Regardless of the placement method you use, the goal is to place the concrete as close to its final destination as possible. Overhandling of the concrete by repeatedly

Place the concrete as close to its final destination as possible. Moving concrete around too much using shovels or other tools can lead to segregation.

dropping or dragging it using shovels or other tools can lead to segregation (when the aggregate separates from the cement paste matrix).

Placement sequence

As a general rule of thumb, start depositing the concrete at the point farthest away from site access. On a driveway, for example, start the placement in one of the corners where the driveway abuts the house or garage and work back toward the street. Place the concrete in strips as squarely as possible. Don't place it diagonally because, in most cases, that's not how the stamp pattern will line up.

Ultimately, you want to be able to place, finish, and stamp the concrete in a methodical, timely fashion. So it's important to sequence the arrival of the concrete trucks accordingly. The number of workers you have on the job and their skill level will often dictate the intervals between trucks. On large projects, most stamping contractors will have separate crews dedicated to placing, finishing, and stamping so the trucks can arrive at more frequent intervals.

Getting Concrete Where You Want It

Decorative stamped concrete is especially popular for pool decks and patios, which means you'll often need to deliver the fresh concrete to the rear of an existing house without damaging the client's lawn, landscaping, or backyard fence.

Obviously, a 10 cubic yard concrete truck is too heavy and unwieldy to maneuver into a small backyard. However, using a wheelbarrow to cart the concrete from the truck can be slow and labor-intensive. Here are some other, more efficient ways to get the concrete where you want it. If you can't justify the cost of purchasing this equipment, you can usually rent it.

Power buggies
These are like large wheelbarrows powered by a motor, allowing you to zip around at speeds topping 5 mph. They are available in walk-behind and ride-on models, and most have capacities of 16 cubic feet.

Skid-steer loaders
You may already own one of these to do excavating work. When equipped with a bucket attachment, you can also use a skid steer to haul fresh concrete. A wide range of bucket sizes and operating-load capacities are available.

Small-line concrete pumps
These compact, trailer-mounted machines can be very handy for residential jobsites where access to forms is difficult. They are small enough to be towed by a pickup truck, and they allow fast placement rates, delivering 25 cubic yards of concrete or more per hour. The ready-mix truck dumps the concrete into the pump's hopper, and then the fresh concrete is sent through a pump line that can extend 500 feet or more (depending on the pump model) to reach the deposit point. Unless you plan to use a pump for a lot of jobs, it's usually more economical to use the services of a pumping contractor rather than buy your own pump. Although I have also used larger overhead boom pumps to place concrete, they are very cost-prohibitive and not commonly used for stamping work. These types of pumps put out massive volumes of concrete quickly (used generally for large commercial pours), which often doesn't allow enough time for stamping work.

(TOP PHOTO) WACKER CORPORATION
(BOTTOM PHOTO) SCHWING AMERICA, INC.

If the ready-mix truck can't access the placement area, use a power buggy (above) or small-line pump (left) to transport the concrete.

Stress to the ready-mix producer in the planning stage of the project that you are placing stamped concrete and the timing of truck arrival will be critical. You don't want the trucks to show up before your crew is ready for the concrete, nor do you want an extended delay so you end up with a cold joint. This occurs when the first concrete pour begins to harden before the next batch of concrete is placed, not allowing the second pour to blend in seamlessly. A cold joint is not only unsightly, it often results in a crack.

As the concrete trucks arrive, you should have a basic understanding of what the

Before placing the concrete, use plastic sheeting to protect adjacent surfaces from concrete splatter.

proper slump of the concrete should be to make sure it is the proper consistency for the job (see How to Take a Slump Test on page 76). I aim to maintain a 4 to 5 inch slump from load to load. If you're using a dry-shake color hardener, anything less than a 4 inch slump will not provide sufficient moisture to wet out the hardener. And generally, concrete with slumps exceeding 5 inches will contain too much water, which can affect the durability of the concrete (as discussed in Chapter 7).

Until you get your sequencing and timing down pat, I recommend that you stick with smaller jobs requiring no more than 5 or 6 cubic yards of concrete at a time. As your experience level and crew size increase, you can take on larger commercial pours, which may require placing, finishing, and stamping as much as 20 or more cubic yards of concrete a day. Regardless of the size of the placement, you should plan beforehand to have enough stamping tools and workers on hand to get the job done in the allotted time frame.

Special considerations during hot weather

A 10° F rise in outdoor air temperatures can increase the initial setting time of the concrete by one-third, really cutting into the time you have available for decorative stamping. If you are unable to keep pace with the faster setting time, consider using step retardation to incrementally slow concrete set.

As discussed in Chapter 6, retarding admixtures suspend concrete set and give you additional working time. Generally, the admixture is added to the concrete at the ready-mix plant in a single dose. But you can opt instead to add the admixture in stages at the jobsite by using a prepackaged product. This will allow you to maintain greater control.

For example, when the concrete first arrives, start by placing one-third of the load (this batch would contain no retarder). Then put the recommended dosage of retarder into the remaining concrete, and place another third of the load. If you need more finishing and stamping time, you can add yet another

TIP

Before placing the concrete, be sure to use plastic sheeting to protect adjacent buildings, landscaping, and existing slabs from concrete splatter as well as any shake-on hardener or powdered release agent you may be using. A little time spent doing this before the pour can pay big dividends later by minimizing cleanup and preventing customer complaints.

Be particularly vigilant about protecting existing stucco or brick walls, because it's extremely hard to remove concrete splatter and stains from these surfaces. Other areas to mask off include pool copings, existing curb and gutter, and concrete borders that you may have poured the previous day.

The ideal slump for stamped concrete is 4 to 5 inches. A slump test can show if the concrete mix is too wet (top) or too dry (bottom).

How to Take a Slump Test

The slump test, which measures the consistency of fresh concrete, is conducted on the jobsite using samples of fresh concrete as it's discharged from the truck mixer. There are many contributing factors that affect the slump of fresh concrete. While the amount of water needs to be considered, other factors such as the amount of coarse versus fine aggregate, temperature and use of chemical admixtures also play a significant role in the slump. More importantly, the slump test tells you whether the ready-mix supplier has provided concrete of the slump required for the job, as specified in the contract documents.

Little equipment is required to take the test. You need a slump cone (a 12 inch tall mold, shaped like an inverted, truncated cone, with a diameter of 8 inches on the bottom side, or base, and a diameter of 4 inches on the top), a 5/8 inch diameter tamping rod, a scoop, and a ruler. You also need a rigid, smooth, nonabsorbent surface, such as a piece of sheet metal or plexiglass, upon which to conduct the test. Then just follow these steps:

1. Dampen the slump cone as well as the nonabsorbent base. Make sure the base is level and won't be exposed to jarring or vibration.

2. Using the scoop, fill the cone with fresh concrete in three layers of equal volume, with the top layer heaped above the cone. (You can paint lines around the outside of the mold, at levels of 2 5/8 inches and 6 1/8 inches from the bottom, to mark depths for equal volumes.)

3. Use the tamping rod to uniformly rod each layer 25 times. Hold the cone firmly in place during rodding. On the bottom or first layer, start near the perimeter of the cone while progressing with vertical strokes spiraling toward the middle. Each of the next two layers should just penetrate into the underlying layers with the strokes.

4. After rodding the last layer, strike off the concrete level with the top of the cone and remove any spilled concrete from around the base.

5. Lift the cone in a smooth, vertical motion in a time of about 5 seconds. Then place the slump cone upside down beside the concrete sample to measure the distance the concrete sample has settled at its center.

ASTM C 143, "Standard Test Method for Slump of Hydraulic Cement Concrete," gives precise procedures for performing the slump test. This document can be downloaded at the ASTM web site (www.astm.org).

Have plastic sheeting on hand to cover your work in case of rain. Support the covering above the concrete so it won't touch the surface.

dose of retarder to the final third of the load before placement.

Rainy day precautions

The summer months also bring an increase in the occurrence of unexpected rainstorms. Certainly, you should avoid placing concrete on days when the forecast indicates that rain is likely. But even when rain isn't predicted, you should be prepared to avoid disaster in the event of a sudden downpour. I keep rolls of plastic sheeting and tarps on the jobsite that I can use to quickly cover the freshly placed concrete if rain threatens. You'll also need some way to support the covering above the concrete so it won't touch the surface. Often, you can use supplies you already have on the jobsite.

For example, I will sometimes take steel dowels and drive them into the ground on either side of the slab forms. Then I'll take a piece of flexible PVC conduit, cut to the appropriate length, and place each open end over the top of the dowels on either side of the form to create an arch over the slab. (You may need to tape additional conduit to the middle of the arch for extra support, so the arch won't flex from side to side.) You can then stretch plastic over the top of this structure to keep out the rain.

Another good support system can be made by sticking steel reinforcing rods or 6 to 8 foot tall 2x4s into 5 gallon buckets filled halfway with concrete (use what you have leftover at the end of a day's pour). If rain arrives, you can quickly erect tents over the concrete using these ready-made supports, which can be carried from jobsite to jobsite.

Downspouts can also be a potential problem if the outlets are aimed toward the concrete placement. If you get a torrential downpour several hours after you finish stamping the concrete, the water exiting the downspout will flood across the newly stamped concrete and erode the surface. Temporarily divert the direction of water flow by attaching a corrugated pipe to the end of the downspout.

To ensure that enough concrete is ahead of the straightedge as it's pulled, have one or two workers rake concrete up against it using shovels or come-alongs.

Keep a level plane of concrete underneath the straightedge to eliminate low spots, or "birdbaths."

CHAPTER 18

STRIKING OFF AND FINISHING THE CONCRETE

The work you do immediately following concrete placement is critical, since this is when you must create the perfect canvas for decorative stamping. When striking off and finishing concrete in preparation for stamping, you want to achieve two primary goals:

- Level the surface to eliminate any high or low spots.

- Bring enough cement paste to the surface to permit a well-defined imprint and to wet out the color hardener (if used).

Because of the importance of these procedures, I recommend that anyone who intends to install decorative concrete professionally get certified by the American Concrete Institute (ACI) as a Concrete Flatwork Finisher & Technician (see Benefits of ACI Certification on page 81). The information here on striking off and finishing concrete is not intended to be a complete guide, but only to provide tips applicable to stamped concrete work.

Tips for strike off

- It takes a lot of finesse to properly strike off the concrete surface and make it as smooth and level as possible, so only experienced crewmembers should do the work.

- I prefer to use an aluminum strike off (or straightedge) rather than a wood 2x4 for the job because it will remain perfectly straight. Wood tends to warp and bow, which can leave ruts or crowns in the concrete surface.

- Use one or two workers to muck (or rake) the concrete up against the straightedge with a shovel or a come-along to ensure that enough concrete is ahead of the straightedge as it's pulled.

- Keep a level plane of concrete underneath the straightedge to eliminate any low spots, or "birdbaths." These depressions in the surface can haunt you throughout the rest of the project. Any color hardener you broadcast onto the slab can settle in these dips and be difficult to work into the surface with a bull float or hand float.

- Whenever possible, strike off the concrete all the way from one edge form to the other. Be sure to get right up to the edge. Stopping a foot or so away from the form can leave a valley there.

- On smaller projects, you can use the tops of the forms as guides for the correct strike-off elevation. If you're striking off a large area, use wet screeds or grade stakes with screed pins to establish finish grade. With wet screeds, a portion of the freshly placed concrete is struck off between two reference points. Then

After striking off the concrete, use a bull float to obtain a level, void-free surface.

> **Removing Excess Bleed Water**
>
> To avoid working in bleed water that's lingering on the surface, here are a few helpful hints for removing it. Stretch a clean garden hose from one side of the slab to the other side, laying the hose flat on the surface. Then drag the hose across the surface, like a squeegee. You can also splice together pieces of PVC conduit and use them in the same fashion. To remove smaller areas of bleed water, try running a wood bull float lightly across the surface. Make sure that the float is flat as opposed to cocked, so it will drag the water off of the surface as opposed to working it into the surface.

the struck surface (or wet screed) is used to guide strike off of the concrete in adjacent sections.

Tips for finishing

- Make sure to use the right tool for the job. For non-air-entrained concrete, I recommend using a laminated wood bull float because it does the best job of smoothing and leveling the surface. The weight of the wood float makes it effective at cutting high spots, filling low-lying areas, and consolidating the aggregates to bring more paste to the surface. However, if the concrete has a high air content (which can make it sticky), a magnesium bull float often works best. I keep both types in my truck. (For more information about finishing tools, see Chapter 27, *Tools, Equipment, and Supplies*.)

- Never float or trowel a surface that has standing bleed water on it. This can weaken the layer of paste at the surface. (See Removing Excess Bleed Water on this page.)

- If you are applying a dry-shake color hardener after initial bull floating, wait until the right stage of plasticity has been achieved. Depending on ambient temperatures, the waiting period can range from 20 minutes to an hour or longer. No bleed water should remain on the surface. (See Chapter 19, *Applying Color Hardener*).

- Initial bull floating should produce a level, void-free surface. Otherwise, you will have difficulty when you use the bull float again to work in the color hardener. If there are low-lying areas or voids, the bull float will pass right over the top and the hardener will need to be worked in with hand floats or trowels. At least 90 percent of the color hardener should be able to be worked in with the bull float.

- Although there's much debate about using steel trowels, or fresnos, to finish exterior flatwork (especially in freeze-thaw climates), they can be beneficial on stamped concrete by producing a smoother, flatter surface prior to the application of the release agent. The main concern is that using a steel trowel could promote premature sealing of the surface, which can trap in bleed water and weaken the surface paste. However, when used properly under the right conditions, trowels can be very effective. Do not begin troweling until well after the last bull float operation and when no moisture is present on the concrete surface (typically one to two hours after concrete placement).

- Under no circumstances use a steel trowel in place of a bull float. A trowel

Begin troweling after the last bull floating operation and when no moisture is present on the concrete surface. Steel trowels with rounded edges are less likely to leave trowel marks in the concrete surface.

will not cut and fill like a float and could prematurely seal the surface. The sequence is always as follows: Apply first application of dry-shake color hardener, bull float, apply second application of dry-shake color hardener, bull float, and when applicable use the trowel. In certain parts of the country it is recommended to not steel trowel exterior concrete, in which case, the final pass would be a magnesium float finish prior to the release application.

AMERICAN CONCRETE INSTITUTE

This is to certify that

JOHN M DOE

has demonstrated knowledge and ability by successfully completing the ACI certification requirements and is hereby recognized as an

ACI CONCRETE FLATWORK FINISHER & TECHNICIAN

Certified Date: 1/1/04   Expires: 1/1/09

Examiner: _____

Benefits of ACI Certification

Simply presenting a portfolio showing impressive photos of your stamped concrete work may not be enough to win a client's business. Most of your prospects will also want reassurance that you are educated in the basics of concrete and are knowledgeable about the proper methods, materials, and tools to use.

I recommend that anyone who intends to install decorative concrete professionally get certified by the American Concrete Institute (ACI) in two areas: as a Concrete Flatwork Finisher & Technician and as a Concrete Field Testing Technician – Grade I. Having these ACI certifications will assure your clients that you know your stuff when it comes to the proper procedures to place, consolidate, finish, joint, cure, and protect concrete flatwork—and that you can verify fresh concrete quality by performing the appropriate field tests.

To receive ACI certification as a concrete flatwork finisher, you must pass a written test and have at least one year (1,500 hours) of ACI-approved work experience, including successfully passing a performance exam. An ACI Concrete Field Testing Technician – Grade I must demonstrate the ability to properly perform and record the results of seven basic ASTM field tests on freshly mixed concrete, including those for temperature, slump, unit weight, yield, air content, and making and curing test specimens. Here, too, you are required to pass ACI written and performance exams.

Once you obtain certification, don't let it expire. ACI requires that you get re-certified every 5 years. (Consider this as an incentive to keep your skills up-to-date.) For more information about these certification programs, visit the ACI web site (www.concrete.org/certification) or call 248-848-3800. Certification training and testing sessions are offered year-round in many cities throughout the U.S.

CHAPTER 19

APPLYING COLOR HARDENER

Dry-shake color hardeners come in a wide array of hues, permitting contractors to produce distinctive effects that complement any architecture. (For a refresher on the various coloring methods, refer back to Chapter 13). What's more, an application of color hardener actually densifies the concrete surface because it contains hard aggregates and portland cement. So just by using color hardener, you are creating a harder, more-abrasion resistant surface.

Possibly the only disadvantage of a color hardener is the effort required to apply it. Color hardeners must be worked into the surface in stages during the finishing process, after initial bull floating. By following the steps presented here, you will achieve good results with minimal headaches.

Though hardeners are often called "dry shakes," they should be broadcast onto the surface with a tossing motion. Work from waist level or lower to keep material from drifting into the air.

> **TIP**
>
> Before broadcasting the color hardener, you may need to "fluff" it first to make it easier to distribute. When containers of hardener sit for a while on a shelf in your shop, the contents tend to compact. If the hardener is packaged in buckets with sealed lids, roll the bucket on a flat surface and then turn it upright before removing the lid. This will help to loosen any compacted material. If the color hardener is in a bag, reach into the bag with gloved hands to loosen the material. When working with color hardener, always wear the proper safety gear, including a dust respirator, nonabsorbent gloves, and safety glasses (see Chapter 11 for additional safety guidance).

Step 1.
Determine how much color hardener you need

Estimating the amount of color hardener you'll need for a given project is simple. In most cases, a 60 pound bucket or bag of hardener will cover approximately 100 square feet of surface area (or about 2/3 pound of material per square foot). So if you're pouring a 1,000 square foot driveway, you would need to order approximately 10 containers of hardener to adequately cover the entire surface. Be aware that lighter colors may require a heavier application, ranging from 90 to 120 pounds (or two containers) of hardener per 100 square feet. Always follow the manufacturer's recommendations for color hardener dosage rates.

Step 2.
Prepare in advance

Have all your color hardener measured out and "staged" on the project before the concrete arrives. Once the pour begins, you don't want to interrupt or delay the finishing operation to open bags and dose color into buckets.

If you're on a large project requiring multiple pours, have the required amount of color hardener for each pour weighed out and ready to go. If you're using several accent colors, be sure to identify the buckets clearly so there's no confusion.

During warm weather, don't keep the color hardener in direct sunlight or in the back of your truck. The portland cement in the hardener retains heat, which could promote faster setting of the concrete surface and lead to crusting. If possible, keep the bags or buckets of hardener in a shaded area, such as under a tree or a building overhang.

Step 3.
Apply the first shake of color hardener

Most manufacturers recommend applying color hardener to the concrete surface in two separate applications, or "shakes." The reasoning is that if you tried to apply the recommended dosage all at once, the application would be so heavy that the color hardener would not have a chance to "wet out," or absorb water. The hardener needs to act as a poultice and draw moisture into it. The usual procedure is to first broadcast two-thirds of the recommended dosage onto the surface and then work it in with a bull float before applying the remaining one-third.

After the concrete reaches the right stage of plasticity (generally when no standing bleed water is on the surface), you can broadcast the first application of color hardener. The wait time can range from 20 minutes to over an hour, depending on the ambient temperature.

Broadcast the hardener by throwing it onto the surface from waist level or

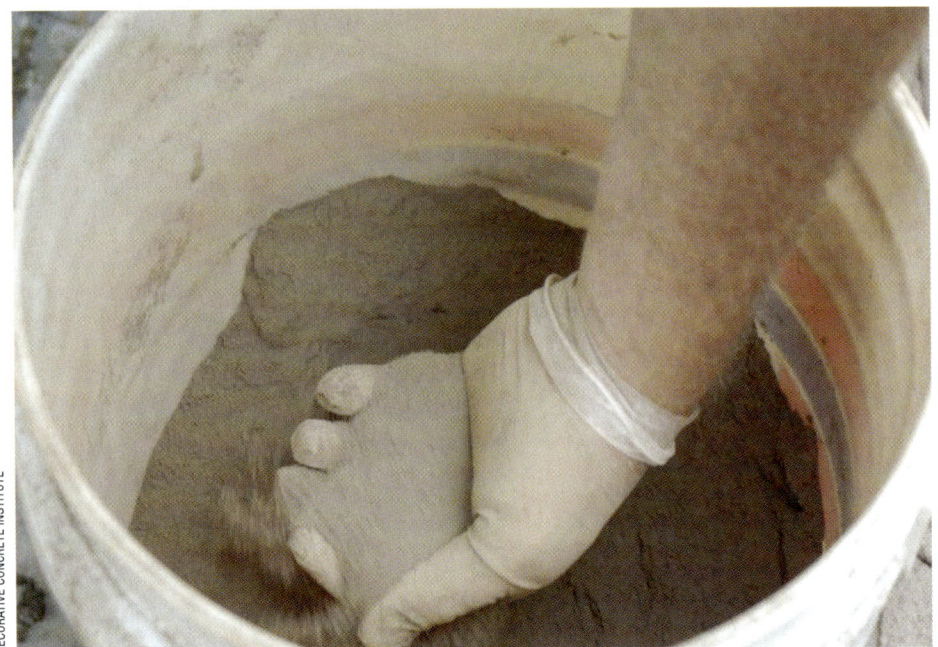

Use a gloved hand to fluff the hardener before application. This will loosen any material that has become compacted during storage and make the hardener easier to broadcast.

a bit lower to minimize the amount of material that drifts into the air. Use a motion similar to throwing a bowling ball, starting with your arm low and behind you and extending it forward with your hand outstretched.

Work from the middle of the slab and back toward the edge forms. This will help to avoid a heavy buildup of color hardener on the edges. To speed application, have one person broadcast from one side of the slab and another person from the other side, with both working from the middle back to their respective edge forms.

Step 4.
Bull float the hardener into the surface

After the first shake of color hardener has been applied to the surface, be patient and give it time to absorb water from the concrete before floating it into the surface.

How to Prevent Moisture Loss

On extremely hot or windy days, you may need to take measures to prevent moisture in the surface from evaporating too fast. Not only can this rapid moisture loss lead to surface crusting and cracking, it will make it impossible for you to properly wet out the color hardener.

An effective way to lock in moisture is to use an evaporation retardant—a waterborne, spray-applied film that temporarily reduces moisture loss. I carry several gallons of it in my truck, just in case I encounter unfavorable weather conditions. If you don't have any evaporation retardant on hand, you can achieve similar results by covering the surface with a sheet of plastic (such as clear 6-mil Visqueen). Avoid at all costs applying water to the surface, sometimes referred to as "water of convenience." You will weaken the concrete by increasing the water-cement ratio.

Apply the evaporation retardant or plastic sheeting right after broadcasting the first application of color hardener. Wait about 10 to 15 minutes until the hardener wets out, and then begin bull floating (if you're using plastic, remove it first). If necessary, repeat this operation after applying the second coat of color hardener.

Don't wait until a state of emergency before using an evaporation retardant. If you anticipate hot or windy weather conditions, apply the retardant right from the start.

Hardener application will go much faster if one person broadcasts from one side of the slab while another person broadcasts from the other side, with both starting from the middle and working back to the edge.

How do you know when the color hardener has wet out sufficiently? Let it sit for 5 to 10 minutes and then, with gloved hands, gently press down on the surface with your fingertips. If you see a fingerprint or moisture where you touched the surface, the hardener has absorbed enough water for floating to begin. Another indicator that the color hardener has wet out properly is a change in color. Regardless of the color of hardener you use, it should darken after absorbing moisture.

If the hardener appears to be absorbing little to no moisture, you could have a potential problem (see How to Prevent Moisture Loss on page 85).

In areas that experience heavy freeze-thaw cycles, the concrete will usually contain an air-entraining agent and a water-reducing admixture. Both can reduce or minimize the amount of bleed water available for absorption by the color hardener. It is imperative to consult with your ready-mix producer and explain that you will need a mix design appropriate for use with a dry-shake hardener.

Step 5.
Apply second application of color hardener

Right after bull floating the first application of color hardener, apply the remaining third of the hardener, repeating Steps 3 and 4. Now you have 100 percent color hardener coverage.

Step 6.
Apply accent, or "flash" colors

For some projects, you may want to apply accent colors of hardener to achieve contrast. This time, your strategy will be somewhat different. Instead of going for uniform coverage, you will use a technique called "flashing" or "flash broadcasting" to achieve subtle color variations, such as you would see in natural stone.

To achieve the most realistic effects, I broadcast a light layer of accent color over the entire surface, making the application slightly heavier in some spots and lighter in others. The goal with flashing is to achieve natural-looking color accents, so be sure to use a light hand. I've seen some horrific flashing jobs with big streaks of color running across the surface. In the case of flashing, less is more.

Once the accent colors have been applied, you are ready for final finishing of the surface with the trowel before you apply the release agent and begin stamping.

After applying the hardener, wait 5 to 10 minutes and then gently press down on the surface with gloved hands. If you see a fingerprint or moisture where you touched the surface, the hardener has absorbed enough water for floating to begin.

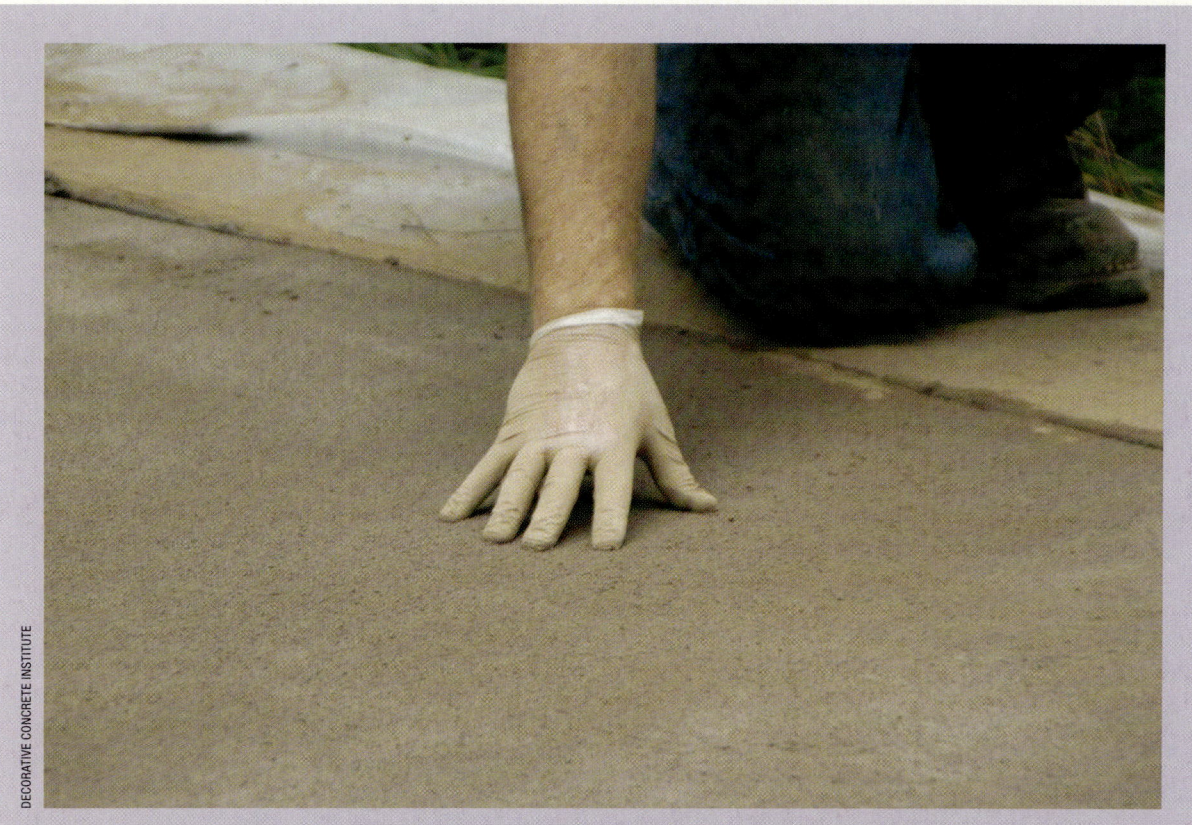

Moisture has been drawn up through the color hardener from where your fingertips were, indicating proper wetting of the color hardener.

CHAPTER 20

APPLYING THE RELEASE AGENT

You've completed the final troweling of the slab. Now it's time to apply the release agent and begin the stamping operation.

As discussed earlier in this guide, powdered release serves two important purposes: It imparts subtle color contrast while acting as a bond breaker to prevent the stamping mats or skins from sticking to the concrete and disturbing the imprint texture. Liquid releases also prevent the mats from sticking to the concrete. Liquid release products are clear when purchased, however if an antiqued appearance is desired, the liquid release can be tinted.

During the planning stage for the project is when you should determine which type of release agent to use: powdered or liquid. Refer back to Chapter 13 for a comparison of these two release types.

Precondition your stamp mats

Whether you decide to use a liquid or a powder, you'll get better results if you precondition, or coat, the stamp mats with some of the release. This will provide additional bond breaking to help ensure a clean imprint.

First, lay the mats—with the textured or

To provide additional bond breaking, precondition the stamp mats with some of the release. If you're using a powder, sprinkle the release onto the bottom of the mats and brush it into all crevices.

patterned side up—on a sheet of plastic. If you're using a liquid release, lightly spray the mats until fully coated. If you're using a powder, sprinkle some of the release onto the bottom of your mats and brush it into all the crevices of the pattern.

Here's another tip that I learned the hard way: Be sure to clean off the stamp mats between uses, especially if you're switching to a different release color. For example, if you use a dark brown release on one job and then a few days later use the same stamps on another job with a light charcoal release, the dark brown could contaminate the charcoal color if you don't thoroughly clean the mats first. Even if you use the same release color from day to day, it's important to clean the mats if you see a layer of paste buildup that could affect the impression on your next stamping project.

Applying powdered release

Most manufacturers package powdered release in 30 pound pails. Be sure to check the product literature for the rate of coverage. Typically, 30 pounds is enough to cover an area of about 1,000 square feet (or an application rate of 3 pounds per 100 square feet).

The best way to apply the release is with a dry tampico brush about 8 inches wide (the same type commonly used to clean finishing tools). Dip the brush into the pail of release and fluff it to load the bristles and coat them evenly. Then take the brush by the handle, holding it below belt level, and use your wrist to flick the release onto the surface in a light, uniform layer. Avoid too much buildup, which could interfere with the imprint texture, especially lighter textures.

When applying release, try not to work more than a few rows ahead of the row being stamped. In windy conditions, any loose release laying on the surface could blow away. Despite the powder release having a water repellent, if left on the surface for an extended period of time, especially in warm conditions, the release can absorb water from the concrete surface, causing it to dry out too quickly during the stamping phase.

> **TIP**
>
> When you begin stamping the areas where release powder has been applied, watch for a lightening in the color of the release to help determine whether you've depressed, or "bottomed out," the stamp completely. Regardless of the color you are using, the release will lighten when it has been fully compacted by the stamp. If you still see some dark areas of release on the surface once the stamp mat has been lifted, that's an indicator that the stamp wasn't depressed to full depth. During the cleaning phase, not only can the release wash away in those spots because it wasn't embedded in the concrete surface, you will end up with insufficient texture from the stamp.

Apply the powdered release with a dry tampico brush, loading the bristles evenly with material. Use your wrist to flick the powder onto the surface in a light, uniform layer.

When applying release, don't work more than a few rows ahead of the row being stamped. The release agent may begin absorbing water from the concrete, causing the surface to dry out too quickly.

Applying liquid release

Using a pump-type sprayer, apply a uniform layer of liquid release onto the surface of the concrete right before you stamp. If you plan to use a tinted liquid release (I discuss how to tint a release in Chapter 13), add the tint a day or two beforehand if possible. This will allow the pigment particles to fully dissolve. Then right before you pour the release into the pump sprayer for application, stir it well (I use a power drill equipped with a paddle mixer) to mix in any release powder that may have settled to the bottom of the bucket. If the release is not thoroughly mixed before you apply it, you could end up with splotchy areas and nonuniform color. Some contractors even strain the tinted release through cheesecloth or a strainer to eliminate any clumps of pigment. Also, shake the sprayer as you apply the release to keep the contents agitated so the pigment doesn't have a chance to settle.

Another word of advice: If you're applying a pigmented release to a steeply sloped area, be especially careful not to flood the area, causing the liquid release to run down the contours of the slope. You may end up with streak marks in the surface when the release dries.

Another way to obtain a random antiquing effect is to apply a very small amount of powdered release to the surface of the concrete and then spray the liquid release over the top of it. The liquid dissolves the light layer of powder to leave subtle accents after the surface is stamped.

Surface Texture Can Affect Release Appearance

When using a powdered release, be aware that the surface texture of the concrete can influence the final look you achieve. During the stamping operation, some of the release gets depressed into the concrete surface to impart permanent color, resulting in an antiquing effect. If you apply the release over a rougher texture, such as a bull floated finish, more release agent will become embedded in the surface during stamping and you won't get as much contrast when you remove the residual release later (see Chapter 24, *Release Removal*). You'll achieve greater variegation when the release is applied to a smoother surface (one that's been finished with a steel trowel or floated with a magnesium float). The smoother finish won't hold as much release and will permit easier removal in the areas where it didn't become embedded by the stamp.

The plasticity of the concrete will also affect the amount of release that becomes embedded in the surface. The sooner you stamp, the greater the amount of release that will combine with the surface paste. If you stamp later, more release will be removed from the surface during the cleaning operation.

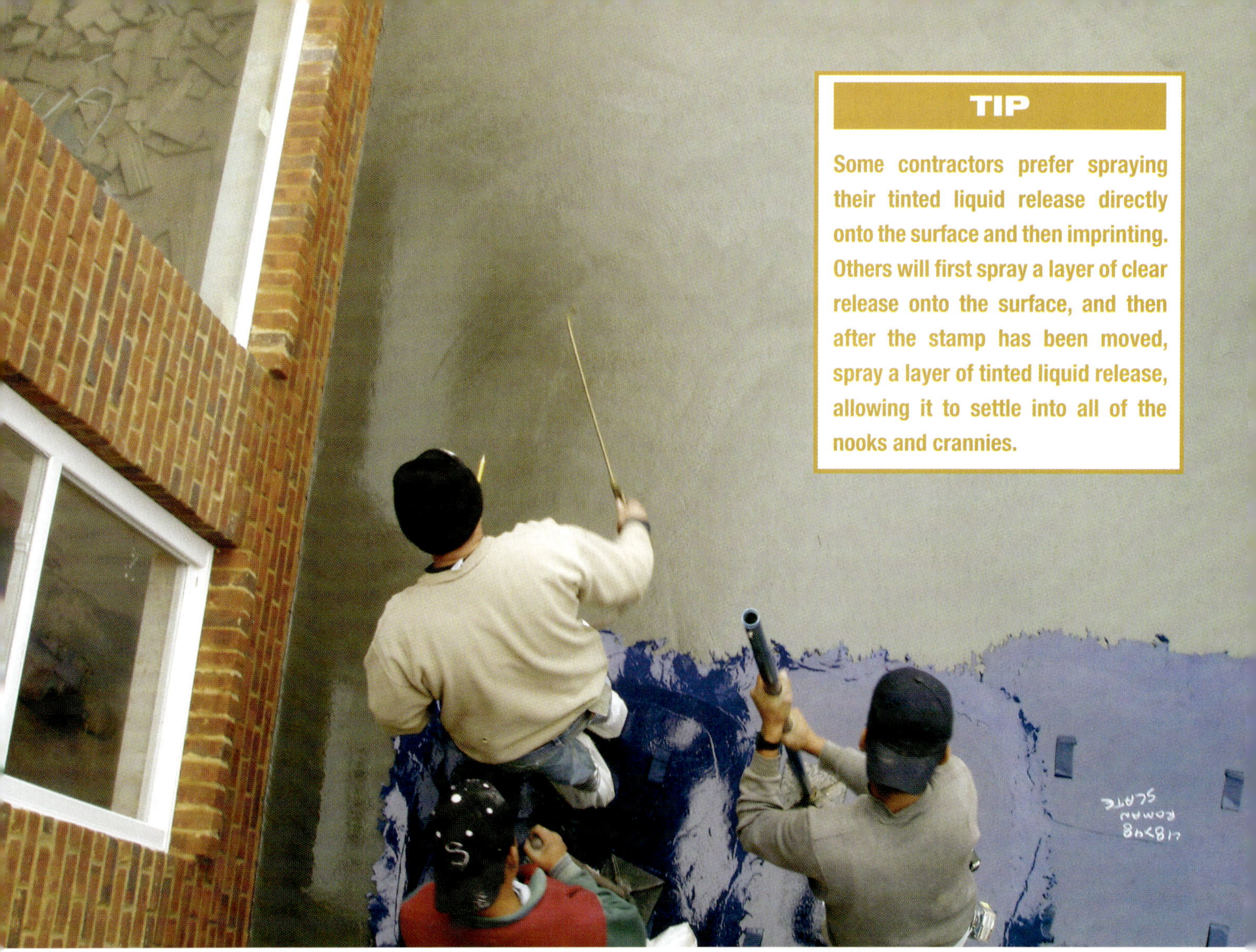

TIP

Some contractors prefer spraying their tinted liquid release directly onto the surface and then imprinting. Others will first spray a layer of clear release onto the surface, and then after the stamp has been moved, spray a layer of tinted liquid release, allowing it to settle into all of the nooks and crannies.

DECORATIVE CONCRETE INSTITUTE

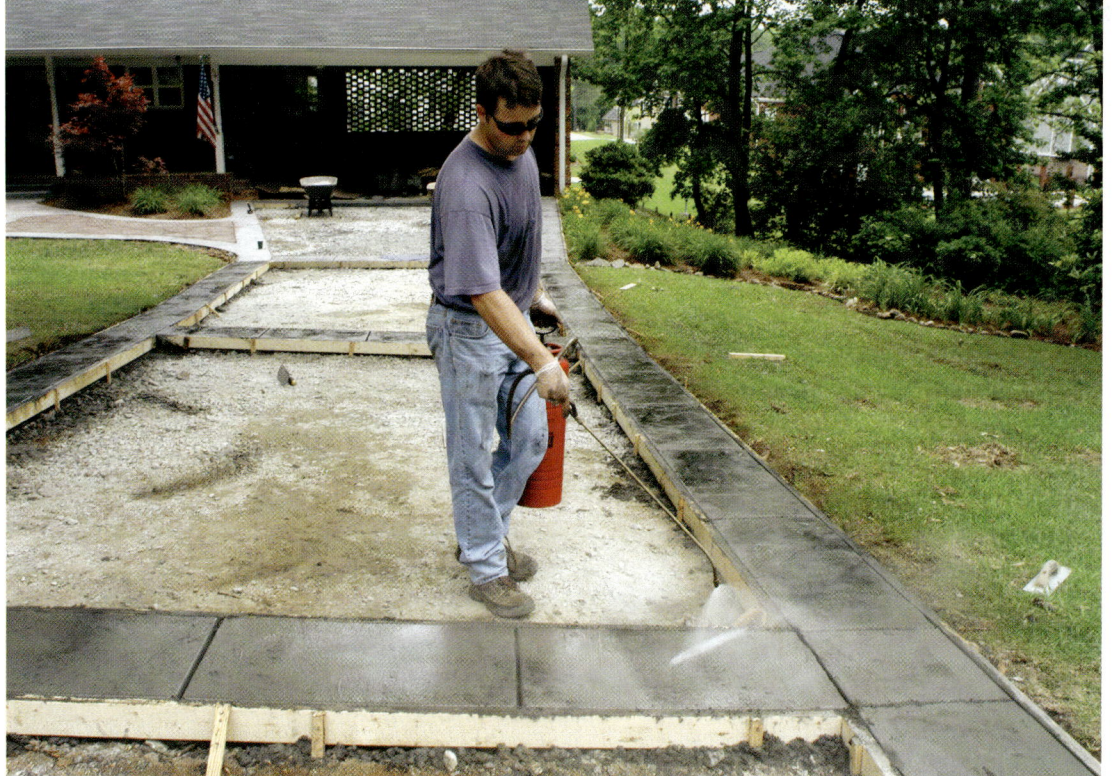

Apply liquid release in a uniform layer using a pump-type sprayer.

A set of mats include several tools that give slightly different impressions of the same pattern to avoid repetition. Some stamp manufacturers use color-coding to differentiate the pattern variations.

Don't forget to pretexture along the perimeter of the slab with a texturing skin or flex mat before you begin stamping. The more rigid stamping mats you'll use for the rest of the slab tend to overlap the forms and won't give you a complete imprint along the edges.

CHAPTER 21

THE CONCRETE STAMPING PROCESS

With the release applied, you can now begin transforming the slab surface using your stamping mats and texturing skins. You generally have a short window of time in which to complete stamping, especially in hot weather, so being organized and well-prepared is a necessity. Diagram your stamping layout in advance, have your stamping tools lined up and ready to go, and make sure you have enough labor on hand for the volume of work being done. If you're using a new stamp pattern for the first time, practice with the tools on compacted sand before using them in concrete (where the results—and your mistakes—could be permanent).

In this chapter, I'll take you step by step through the stamping process. But first, let's discuss some of the factors to consider when choosing stamps.

Selecting your tools

With the growing popularity of decorative stamped concrete, manufacturers have responded by offering an ever-expanding selection of stamping tools and patterns. (You'll find some of these suppliers listed in the resources section on page 138.)

Most stamping mats today are constructed of a durable polyurethane that provides many reuses. Both rigid and semi-rigid types are available. Some contractors prefer rigid mats, claiming that they allow workers to start the stamping process sooner. Others swear by a more flexible mat, saying they can "feel" the underlying contour of the concrete better. Also, a flexible mat may work better where there are undulations in the slab, such as on sloped areas or on driveway aprons where the ends flare up slightly.

A highly accomplished stamper can generally work with both types of tools, taking advantage of the positive attributes of each. Do your homework and ask experienced stampers about the tools they prefer to use and why. Also, call several stamp manufacturers and ask them to explain what makes their tools superior to others.

Regardless of the stamps you choose, make sure you purchase enough tools for the job. This is critical to doing the work successfully (see How Many Stamps—and Workers—Do You Need? on page 94).

Generally, you'll buy stamps in sets consisting of several mats giving slightly different impressions of the same pattern. The reason the stamps aren't identical is to avoid pattern repetition and produce a random look that's more realistic. Some mats are marked with letters (such as A, B, C) to differentiate the pattern variations, while others use numbering systems or color-coding.

As a companion to your set of stamps, purchase at least one flex mat, or floppy, having the same texture and pattern. A floppy is more flexible than a standard mat and will bend easily when you're working up against walls, columns, and other vertical surfaces. You should also purchase several texturing skins to texture along the perimeter of the forms and vertical faces, such as step risers. Skins are thin and pliable and add texture only. They are also great for fixing blemishes from nonuniform stamping.

When to start stamping

Before you begin, check to see that the concrete has reached the right stage of plasticity. If you begin stamping too soon, the concrete won't be firm enough to support the weight of workers or hold a well-defined imprint. If you start stamping too late, not only will stamping require more effort, you'll produce little or no texture with the stamps, especially as you reach the end of the job. I have suffered my share of blisters from stamping too late in the process. Timing is everything!

I don't use any sophisticated methods to determine the right time to start stamping. I simply press my fingers into the concrete surface at several locations on the slab. If I leave a clean imprint about 3/16 to 1/4 inch deep, without exposing the aggregate or seeing surface moisture, I know I can begin stamping with most patterns. Note: Light textures may require you to wait longer.

For the average job, three people should be devoted to the stamping work so their progress won't be interrupted. Have other crewmembers on hand for miscellaneous tasks.

How Many Stamps—and Workers—Do You Need?

Having enough stamps is crucial on any decorative stamping project. Don't ever think you can tackle a job using only two or three stamps. With stamping tools, more is always better.

As a general rule, you should have at least enough tools to form a row spanning the widest area of the slab plus two extra stamps to start another row. For example, if you are using 24x24 inch stamping mats for a residential driveway 20 feet wide, you would need 10 mats to span the 20 foot width, plus two additional tools. With some irregular notched patterns, it may be more efficient to lay out more than one row at a time, which means you'll need even more stamps, especially on larger projects.

If you plan on doing a lot of commercial work involving large slabs, consider doubling the number of stamping tools you purchase so you can cover large areas expeditiously. I have a friend in Philadelphia who stamps over 500,000 square feet a year. He purchases as many as 20 or 30 stamps per set so his crew can work on three or four rows at a time.

That brings up the topic of crew size. How many workers will you need? Of course, the answer depends on the square footage and complexity of the project. But for the average job, I like to have at least five workers on hand, with three people doing the stamping and two additional people for other tasks, such as placing and texturing borders. The crew doing the stamping should not stop the process once they get started.

Weather conditions can also be a factor. During hot weather, you need to move across the surface quickly, before the concrete gets too firm to stamp properly (unless you've taken efforts to control concrete set, as discussed in Chapter 6). Having more tools and additional workers on hand for tamping can speed the process.

Another test: The stamp should hold your weight when you step on it and not slide around or sink too deeply into the surface. If the concrete is just right, standing on the stamp should be all it takes to create an impression. Generally, the deeper the stamp texture or pattern, the sooner you can start the stamping process.

Pretexturing the slab perimeter

The first step in the stamping process is to pretexture along the perimeter edges of the slab with a texturing skin or flex mat. What you're trying to do is get texture into the concrete and press the release into the surface from the edge form to approximately 6 to 12 inches inward.

This step is important, because when you're working with a nonflexible stamp, the tool will overlap the edge of the form and you won't be able to fully depress it into the concrete surface. By pretexturing the perimeter first, you'll get the texture you need and the full color from the release. Remember, if the release agent doesn't get compressed by the stamp, it

If the concrete is at the right stage for stamping, you should be able to impress the stamps into the surface by simply walking on the tools, possibly followed by a light tamping

When picking up and moving the stamps, be on the lookout for any problems or imperfections, such as stamps that haven't been tamped down with enough pressure to leave a complete imprint.

will simply wash off the surface without leaving the desired color (as discussed in Chapter 20).

The stamping sequence

Once the edges are pretextured, your crew can begin stamping the rest of the slab with the mat tools. This should not be done haphazardly, however. It's important to do the work in the proper sequence.

Generally, you should stamp in the same sequence that you placed and finished the concrete. For example, if you started placing the concrete in the top left-hand corner of the slab and ended on the bottom right corner, this would be the preferred sequence to use for finishing and stamping operations, working row by row from the starting point to the end point. The reason is simple: The concrete placed first will reach the right stages of plasticity sooner.

You may run into some exceptions, however, such as an area in the middle of the slab that's

After setting the first row of stamps, step back and eyeball them to make sure they're running square and true. If you have doubts, check the alignment using a string line or tape measure.

Never Trowel the Release

An important thing to keep in mind while stamping is that a release is a bond breaker and coloring agent. It serves no structural function, and it should never be troweled into the cement paste because it could affect the structural integrity of the concrete surface and cause discoloration.

If you need to trowel and restamp a problem area, such as where the stamps have drifted or someone has stepped onto the surface, you must remove all of the release first, whether it's a powder or liquid. With a powdered release, brush it off with a stiff-bristled broom until you expose the underlying color. Once the release is completely removed from the affected area, you can then trowel, reapply the release, and continue stamping. With liquid release, try to absorb as much of the liquid as possible with a rag or sponge before troweling.

shaded by a tree. In that case, the stampers may need to work around the shaded area if it hasn't reached the proper stage of set when they get to it. Going back to stamp problem areas requires some skill, because you must align the stamp pattern with the surrounding areas that have previously been stamped. You'll learn with experience how to adjust the stamping sequence as needed.

Another objective when stamping is to avoid pattern repetition, especially with patterns that mimic natural materials such as stone or slate. A random composition will look much more realistic. That's why most stamp sets are labeled with letters or numbers, as mentioned earlier. Always arrange the stamps in the sequence recommended by the manufacturer, such as A,B,C or 1,2,3. Never set an A stamp against another A or a No. 1 against another No. 1, for example.

Checking stamp alignment

It's imperative to establish the first row of stamps on a straight line because that will serve as your benchmark for the rest of the job. For example, when stamping a driveway, don't use the edge of the garage floor as a benchmark unless you've determined that the edge is perfectly straight. If it's not, the rest of the rows will be out of alignment as well. (If there is a bow or curve in the garage floor, you can compensate for it by overlapping the stamps into the garage as needed to form a straight row.) It's also vital that the forms are squared at the corners (see Chapter 15, *Erecting the Forms*).

After setting the first row of stamps, step back and look at them from every angle to make sure they're running square and true. If you have any doubts, get out your string line. This can be your best friend during the stamping process, especially for aligning stamp patterns that are perfectly square or rectangular. If you need to get out on the slab a bit sooner than usual to cover a large area, the stamping tools may slide or shift on the surface. It's not unrealistic to have them shift as much as 1/4 inch per row, especially on sloped areas. So be sure to periodically check alignment every couple of rows using the string line.

For notched or irregular stamp patterns, you can check alignment by using the edge form as a reference point (assuming that it's square). Use a string line or tape measure and run it from the edge of the form to the top and bottom edges of the stamp mat to make sure the tool is running square relative to the form. It's especially important to do this when stamping the first row. After that, you can check the tools periodically for alignment.

As beneficial as string lines and tape measures can be to check stamp alignment, you should still eyeball your work. Occasionally stand back and view the rows from different perspectives.

A word of caution: Some stampers will snap the string line into the powdered release and use that as a guideline for placement of the stamp edge. That technique works fine if the stamping tool is square or rectangular. If you're using a notched stamp, however, portions of that line will be visible in the finished surface, marring the appearance.

There are several tricks you can use to apply color and texture to vertical faces, such as stair risers.

The stamping process

Now you're ready to stamp the first row. It's best to have enough stamps on hand to finish a complete row plus two additional tools to start the next one. If the concrete is at the ideal stage for stamping, you should be able to impress the stamps into the surface by simply walking on the tools, possibly followed by a light tamping. If you find that at this stage of the project you have to tamp the tools aggressively, you are probably behind the game and started stamping too late. On small projects (using 2 cubic yards or less of concrete), you may get away with waiting longer to stamp, but you'll have to begin tamping from the start. This is not realistic on most projects, unless you have 15 or more mats with multiple people available for tamping.

The stamping crew should complete the first row before moving on to the second one. Typically, one person will place the starter tools and stand on them while grabbing stamps from the first row and leapfrogging them into the next row. While this person is moving and advancing the tools, another person can do the tamping. Depending on the stamp pattern, a third person may be needed to detail the grout joints (more about this later). If all is going smoothly, stamping is typically a speedy process.

On many jobs, you'll be faced with having to stamp up against a vertical surface, such as a wall or a column. This is where floppy mats and texturing skins come into play. They can bend or flex up against these surfaces, unlike the more rigid tools.

Usually it's the responsibility of the person who is picking up and moving the stamps to be the observer, looking for any problems or imperfections. For example, if he notices that the stamps haven't been tamped down with sufficient pressure for a full imprint, he should tell the person doing the tamping to use greater force. (Refer back to Chapter 20, *Applying the Release Agent*, for tips on how to tell if you're getting full texture from the stamp.) It's best to fix any imperfections as soon as possible. Once the crew gets halfway down the slab, it may be too late to correct problems. That's why it's also important to check stamp alignment every few rows.

As the work proceeds, watch that moisture doesn't begin to accumulate on the bottom of the stamps. This can happen during humid weather conditions or when you're working on a concrete surface that's overly wet. If you see moisture beginning to collect, stop using that tool immediately. If you continue to use it without removing the moisture, the tool will pick up more and more paste as you move along, and you will see those imperfections all the way down the line. If you're using a powdered release, you can remove the moisture from the bottom of the stamp by brushing it with some of the release powder. If you're using a liquid release, remove the moisture with a dry rag or a sponge before putting the tool back into service.

As the stamps are being advanced, use a hand roller to touch up joints if necessary.

Also be sure that the person walking on and moving the stamps is wearing clean boots or work shoes, free of any pebbles, mud, or other debris. Inevitably, these contaminants will end up on top of the stamps and fall onto the fresh concrete surface as the stamps are being lifted.

Detailing

It will pay big dividends to have a good detail person on the job—someone who follows behind the stamping process to fix minor blemishes and dress up the grout joints (in patterns that have them). Even if you've pretextured the edges and used a flex mat against walls, you will often find it necessary to do some detailing with a hand chisel, roller, or texture skin.

Here are some common situations requiring detail work:

- To remove "squeeze"—displaced cement paste that comes up through the grout joint between the stamps. This is especially common with patterns that have very wide or deep joints.
- To fix blurred pattern lines caused by double stamping (where the stamp slightly overlaps the edge of a previously made imprint).
- To correct grout joints where the stamp wasn't tamped down with sufficient pressure.
- To hand chisel pattern lines where the stamping tool has not made a full imprint, such as around perimeters.

With most stamp patterns, you'll achieve better results if you detail the same day, either as the stamps are being advanced or before going home at the end of the day. As a last resort, chisel the joints the following day.

To remove "squeeze" between stamps, it's usually best to roll out the displaced cement right away, especially with wider grout joints. This will result in a more natural looking joint. If the joint is narrow, however, rolling it immediately after the stamp has been moved may actually make the joint look worse by widening it. In such situations, you're often better off waiting until the next day, when you can usually snap away the thin vein of paste.

If you wait until later in the day to detail, be careful not to disturb the imprinted surface when walking out onto the slab. You can either stand on texture skins, leapfrogging them as you go, or wrap your feet with some Visqueen to create a plastic "pillow."

For touching up or fixing minor surface flaws, a texture skin can be your biggest ally. It can be used as an eraser to correct unevenness or nonuniformity. For example, sometimes a stamp may get pounded into the surface too deeply. When you remove the adjacent stamps, you will end up with high and low spots. Immediately use the texture skin to hand pat these areas until they are level and then restamp with the appropriate mat tool.

Vertical stamping

On some projects, you may need to apply pattern and texture to vertical faces, such as stair risers. As discussed in Chapter 15, the key to doing this is to strip the forms while the concrete is still plastic enough to take an impression. You can then apply a flex mat or texturing skin onto the vertical face and use a hand tool to cut grout lines where necessary.

If you want to apply a dry-shake hardener for color, there are several ways you can work it into the vertical faces before texturing. One method is to use a rubber float to work up some paste immediately after the vertical face has been stripped. Once enough paste has been worked to the surface, you can then take a handful of the color hardener and throw it onto the vertical surface. The color hardener should stick to the paste and absorb enough moisture to wet out. A small float or trowel can then be used to work the color hardener into the surface.

Another method is to mix up a paste of the color hardener with water and then trowel the paste onto the surface. If using this method, apply the paste shortly after the vertical face has been stripped so that it can bond and harden cohesively. Don't apply the paste too thickly (no greater than 1/4 inch) or it could shrink and produce cracking.

Another method I use is called "porridging." This works well for cantilevered steps. Rather than stripping the forms while the concrete is still plastic, and risk breakage of the cantilevered edge, I leave them in place until the next day. Then I mix up a "porridge" of color hardener, combining it with a bonding agent and water, and plaster this mixture onto the riser faces. Predampen the area with this bonding agent/water solution by brushing it onto the surface right before applying the color hardener paste. Before the paste hardens, apply the same type of release agent used for the rest of the job, whether a liquid or powder, and then apply the texture skin and hand tool any grout lines.

BRICKFORM RAFCO PRODUCTS

DISTINCTIVE CONCRETE

CHAPTER 22

CURING STAMPED CONCRETE

Once the stamping process has been completed for the day, your next concern should be curing the concrete. The goal when curing is to retain sufficient moisture content for a long enough time to allow the necessary properties of the concrete to develop. With proper curing, concrete becomes denser and less permeable, resulting in an overall increase in strength and durability.

Although it's generally recommended to start curing concrete as soon as possible to get the maximum benefits, there are some variables that will affect which method you use and the timing of curing. Following are some general guidelines I go by. For more detailed information on curing procedures and materials, a good resource is the American Concrete Institute's *Guide to Curing Concrete* (ACI 308R).

Powdered vs. liquid release

If you've applied a colored release powder to the concrete surface, you can't apply a curing compound until you wash off the residual release agent—a minimum of one day and in some cases two or three days later, depending on weather conditions. (For more information on procedures for release removal, read Chapter 24.) Once the surface is sufficiently cleaned and allowed to dry, you can then spray on a liquid membrane-forming curing compound or a cure and seal to retain sufficient moisture in the concrete (see What Is a 'Cure and Seal'? on page 102). Make sure you apply these products at the manufacturer's suggested coverage rates.

If you're using a clear or tinted liquid release, you can usually apply the curing membrane to the slab the same day, assuming there's no residual release remaining on the surface. Check the release manufacturer's recommendations for curing.

Once the slab has cured sufficiently, you should apply a finish coat of sealer. Most manufacturers recommend applying the sealer several weeks later, after a light surface cleaning. Be careful not to apply the sealer too heavily, which could trap moisture in the slab. See Chapter 26, *Sealing Stamped Concrete*, for more guidance on proper sealer application.

Proper curing helps concrete become denser and less permeable, which contributes to an overall increase in strength and durabilty.

Extreme weather conditions

The need for adequate curing is even more important when you're placing concrete in extreme weather conditions. If the weather is hot or windy with low humidity, the concrete will rapidly lose moisture through evaporation. Covering the concrete with a layer of impervious building or Kraft paper can be an efficient means of preventing this moisture loss. The paper should be applied before your crew goes home for the day but not so soon that marring of the surface could occur. Generally, if the surface is hard enough to walk on, it's ready for application of the curing cover.

Do not tape the seams of the building paper with duct tape. Instead, lightly weigh down the paper with straw or a thin layer of sand. The duct tape does not allow moisture to pass through and can actually retard the layer of surface paste directly underneath the tape, resulting in uneven surface hardening. Unfortunately, I learned this lesson the hard way. On one job, when my crew and I came back to remove the paper and wash the slab (thinking it had hardened sufficiently), we ended up removing strips of the surface layer precisely where the duct tape was placed.

Plastic sheeting is also a very effective way to prevent rapid moisture loss in concrete slabs. However, when placed directly onto stamped surfaces, the plastic can leave permanent crease lines and cause uneven coloring. Because plastic is difficult to pull tight with no creases, I first put down a layer of building paper and then cover it with plastic.

If you're placing the stamped concrete in cooler weather, you should cover the concrete with insulating blankets, especially if the temperatures might drop below freezing at night. Freshly placed concrete is especially vulnerable to freezing because it's in a saturated condition. The blankets will help to retain the moisture and heat needed for proper curing.

Because local weather conditions vary considerably, I suggest asking your ready-mix concrete supplier or local ACI chapter for tips on the best curing procedures to use in your area if extreme temperatures are anticipated.

What Is a 'Cure and Seal'?

Traditionally, concrete curing and sealing take place in two separate steps. As the name implies, a cure and seal product will both cure and seal the concrete in one application. Most cure and seals are water- or solvent-based acrylic resins. Upon drying, they form a membrane that retards or reduces moisture evaporation from the concrete surface while providing protection from abrasion, staining, and harmful chemicals, such as deicing salts.

Be sure to use a cure and seal product that meets the requirements of ASTM C1315, *Standard Specification for Liquid Membrane-Forming Compounds Having Special Properties for Curing and Sealing Concrete*. Also, closely follow the manufacturer's recommendations for application rates.

Applying a liquid curing compound or a cure and seal product to freshly stamped concrete reduces moisture evaporation, giving the concrete time to gain strength and durability. Closely follow the manufacturer's recommendations for the timing of application and coverage rates.

CHAPTER 23

INSTALLING JOINTS

After your meticulous efforts to imprint a pattern or design in concrete, you'll want to do everything possible to prevent conspicuous cracks from ruining the beautiful surface your client paid for. Although you may not be able to prevent all cracking in concrete due to the stress caused by temperature changes and drying shrinkage, you can be proactive about where the cracks occur.

Cutting contraction joints (also called control joints) at the proper depth and spacing in the slab soon after placement provides stress relief at planned locations and prevents uncontrolled random cracking. If the concrete does crack, it will occur right beneath the joints and not be visible on the surface. Even slabs with deep-cut stamped patterns must be jointed to control cracking. But you can often hide the joints by integrating them into the pattern lines.

Even if you install the required joints, however, they will do little good unless they are at the correct spacing and cut to the proper depth.

A groover with a V-shaped bit is used to form joints in the concrete before it sets.

All concrete slabs require contraction joints to prevent uncontrolled random cracking. With stamped concrete, however, you can make the joints less noticeable by integrating them into the pattern lines.

For more information

The following publications from the American Concrete Institute are excellent references for information on joint design and construction as well as load transfer at joints:

- *Guide for Concrete Floor and Slab Construction* (ACI 302.1R)
- *Joints in Concrete Construction* (ACI 224.3R)
- *Slabs on Grade* (Concrete Craftsman Series 1)

How far apart?

In general, space joints a distance of two to three times (in feet) the thickness of the slab (in inches). In a 4 inch thick slab, for example, space joints 8 to 12 feet apart.

In addition to slab thickness, other factors can influence how far apart to space joints. The following situations may require more frequent joint spacings:

- Use of a concrete mix with a greater potential for shrinkage. The quantity of cement is the biggest

Snapping a chalk line is a fast way to mark joint lines for sawcutting. A string line can be used to pop lines in fresh concrete for grooving.

factor, with higher contents leading to more shrinkage. Another factor is the nominal aggregate size. A mix with a 3/8 inch pea rock will typically shrink more than one with a larger 3/4 inch rock. Consult with your ready-mix producer.

- Lack of steel reinforcement. For many slabs on grade, steel reinforcement is necessary to add strength and help to control cracking (see Chapter 16). If reinforcement is not specified, the concrete may require more joints to provide crack control.

How deep?

Contraction joints should be cut to a depth of at least one-fourth the slab thickness. For example, in a 4 inch thick slab, cut the joints at least 1 inch deep.

Installation methods

There are several methods for installing contraction joints, but the most commonly used are grooving and sawcutting.

With grooving, you form joints in the concrete as it begins to set using a groover—a hand-held tool with a V-shaped bit made of bronze or stainless steel. (Be sure to choose a bit with a depth at least one-fourth the slab thickness.) With sawcutting, on the other hand, you wait to cut the joint until after the concrete has set, using a saw equipped with a diamond or abrasive blade. A sawed joint is less noticeable than grooving and is the method I prefer.

One of the most popular tools for cutting joints is a handheld cut-off saw, or quick-cut saw. Although some of these saws have an attachment for connection to a water source, most contractors prefer cutting dry. Keep in mind that dry-cut sawing creates a lot of dust, so wearing proper safety gear, such as a dust mask and goggles, is essential. A walk-behind saw can also be used to cut joints in slabs that are imprinted with light to moderate textures. However, on deeply textured patterns with lots of highs and lows, this type of saw is usually not the best choice because the wheels will not ride level on the surface.

If you're using a powdered release, you can save time by sawcutting the contraction joints before removing the release. This allows you to remove the release residue and the dust created from sawing in one step (see Chapter 24, *Release Removal*). This approach is much more efficient than cleaning off the release, sawing the joints, and then cleaning the slab again.

Other methods for forming joints include installing strips of wood or metal (sometimes called "keyways") or chiseling. For the latter method, I've seen some stamping contractors use custom-fabricated chisels about 3 feet in length to hammer out joints in the concrete. Others use special tools fabricated out of sheet metal and attached to handles to cut the contraction joints as they are stamping. I have had success with this method in the past, but I don't recommend trying it for the first time on an actual project.

A handheld cut-off makes quick work of joint cutting. Another plus: A sawed joint will be less noticeable than a grooved one.

Some contractors use special jointing techniques, such as hammering out the joints with custom-fabricated chisels.

Experiment on a sample until you perfect the technique.

When cutting joints, make sure they are continuous and not staggered or offset, because cracking is likely at T-intersections. Also be sure to put joints at reentrant corners, a vulnerable spot for cracking. A jointing plan can help you predetermine joint spacings and locations (see Start with a Plan on page 109).

When to cut joints

In hot weather, concrete might crack if joints are not cut within approximately four hours after concrete finishing. In cold weather, the window may be much longer. Experience with local weather conditions and mix designs will help you determine the best timing.

Generally, if more than 12 hours have elapsed before sawcutting, you've waited too long and the concrete is already starting to experience shrinkage cracking. If you must wait until the day after concrete placement to cut joints, be sure to start the work first thing in the morning to relieve the stresses in the concrete as soon as possible.

If sawcutting your joints, you don't want to start cutting the joints too soon because the aggregate will tend to break loose and you'll get spalling at the joint. If you see that the edges of the sawcut are fraying and aggregate is being dislodged, wait a bit longer before continuing the job.

Construction and isolation joints

In addition to contraction joints, your stamped concrete slab may require two other types of joints: construction and isolation. Both types run the full depth of the slab and are installed during concrete placement.

Construction joints. If you're placing stamped concrete for a large commercial project, you are probably going to install the slab in multiple pours spanning several days. These jobs will require the use of construction joints, which are placed in the slab where concreting ends for the day (by stopping the pour at a side form).

Construction joints facilitate placement of large concrete slabs by allowing the concrete to be poured in sections of manageable size. However, if the slab will be exposed to traffic loads, you must provide some type of load transfer at the joint to prevent differential slab movement. When construction joints are subject to applied loads that move from one side of the joint to the other, faulting can occur unless adequate load transfer is provided.

I've had good success using slip dowels (smooth steel bars) to tie the previous day's pour into the new pour. The dowels keep the two sides of the joint at the same elevation without restraining horizontal movement, allowing the slab to expand and contract at the joint. Dowels should also be used if you prepour concrete borders one day and then come back a day or two later to place and stamp the infield portion of the slab.

Drill holes at midlevel in the form and insert the dowels in the proper position before the concrete is poured. If you come back after the fact and try to drill holes for the dowels in the hardened concrete using a rotary hammer, you risk fracturing the edges of the concrete.

To function properly, the dowels must be aligned perpendicular to the joint and parallel to the slab surface. They also must be "debonded" from the slab (free to slide) on at least one side of the joint. Some contractors

apply grease to one end of the dowel so the fresh concrete will not adhere to the steel, permitting slight horizontal movement as the concrete shrinks.

Isolation joints. On jobs where the stamped concrete slab will be abutting other building elements such as walls, columns, and foundations, isolation joints are needed to allow the elements to move independently of the slab. Unlike construction joints, an isolation joint must allow both horizontal and vertical slab movement to be effective.

The most common method for creating isolation joints is the use of a preformed strip made of an asphaltic material. This strip is nailed into place at the proper elevation. A foam expansion joint also works well, especially for areas with a radius. The expansion joint material most commonly comes in a 1/2 inch thickness, but a 1/4 inch thickness is also available and will be less noticeable.

Hiding joints

On stamped concrete patterns with grout lines (such as brick and cobblestone designs) you can often hide the joints by integrating them into the pattern. For example, if you plan to space your contraction joints about 10 feet apart, you may be able to adjust the joint spacing 2 to 4 inches one way or the other so the joints will fall where the repeating grout lines have been formed by the stamping tool.

With random stamp patterns, hiding the joint lines may not be possible. Explain to your client that even though some joints may be visible in the slab, they will be much less obvious than the random cracking they help to prevent.

Start with a Plan

Whatever method you choose to install contraction joints, predetermine where the joints will go before the concrete hits the ground. Don't try to figure out a plan during the heat of the pour.

A good jointing plan should clearly show the locations of all contraction, construction, and isolation joints. In the case of contraction joints, the plan should also show which joints need to be cut first and the depth of cut. Start with a scale drawing of the entire concrete slab, representing its exact shape and dimensions. Then draw on the plan with a marker to identify joint locations and spacings (use a different marker color for each joint type). Workers can then use this plan to mark joint locations on the slab forms. After the concrete placement, these guides will show exactly where to cut the joints.

If you plan to use a groover to form joints, snap a string line onto the fresh concrete for alignment and then use a straightedge to help guide the groover as you tool the joint. A chalk line is also a good method for marking joint lines for sawcutting. It is a good idea to try and use orange fluorescent or light colored chalk. Sometimes it is difficult to remove blue or red chalk, especially if the chalk line is too wide and extends outside the boundaries of the cut.

CHAPTER 24

RELEASE REMOVAL

A release agent acts as a bond breaker, creating a nonstick membrane that prevents the stamping tool from sticking to the concrete surface. However, this bond breaking effect can be a detriment later. Unless you remove the residual release (the material that doesn't become embedded in the concrete during stamping), it can prevent the sealer from adhering to the surface. It can also impede the penetration of any stains you might apply later for accent color.

That's why it's important to be meticulous about powdered release removal, cleaning off as much residual as possible without damaging the concrete surface. Typically, 70 to 80 percent of the release is washed away, leaving subtle, permanent color tones. If you use a release powder in a color that contrasts with the underlying integral or dry-shake color, you will achieve a variegated look.

This chapter focuses on methods for removing powdered release agents. If you're using a liquid release, some manufacturers claim that removal usually is unnecessary,

Use a push broom, water, and a mild detergent or cleaning solution to scrub away residual release powder.

If you remove the release too soon, you could damage the concrete surface. Test surface hardness by wetting a small, inconspicuous area of the slab and then gently scrubbing the surface with a brush.

since most of these products will simply evaporate. However, if you've applied the liquid release too heavily, especially if it's tinted, a light cleaning of the surface is recommended prior to sealing.

When to remove the release

Don't start removing the release until you can do so without damaging the surface. That means allowing the concrete to reach sufficient hardness so that the top layer of cement paste isn't removed right along with the release. In cooler weather, you may need to wait three or more days for the surface to harden. In warmer weather, you may be able to start removing the release the next day.

Be cautious and test the surface hardness before you begin. In an inconspicuous area of the slab, clean a small spot by wetting it and gently scrubbing with a hand brush. If the surface paste has not reached sufficient strength, the cleaning process will begin to erode away not only the color imparted by the release, but in some cases the color hardened or integrally colored surface as well. The surface will also appear grainy because the fine aggregate in the paste is being exposed. If you experience this, let the surface harden at least one more day and possibly longer. Then try the cleaning process again, starting out slowly and nonaggressively until you're sure the surface has reached the proper strength.

If you plan to sawcut the contraction joints, I recommend doing the job before release removal. Sawcutting creates a lot of dust, so it makes sense to clean away this dust residue at the same time you're removing the release, rather than cleaning the slab twice.

Removal methods

Some release manufacturers recommend using a pressure washer to remove the release. While this method may be fast, I'm not a fan of using it. I've seen several jobs that were ruined by contractors who used pressure washing too aggressively or too soon. Pressure washing can sometimes remove too much release and strip away some of the color it imparts to the surface, making it necessary to restore the color later (see Chapter 25, *Fixing Minor Flaws*). With pressure washing, it's also hard to control water runoff. If you're working in a residential community, for example, you don't want to cause a mess by washing off the release and allowing the runoff to go into the street and/or into someone's landscaping.

First, I try to sweep up as much of the dry release as possible with a push broom. This helps to minimize the runoff once water is applied. And because the dry release can act as a water-repellant, the water may just bounce off the surface if too much is remaining. Once I remove the dry release, I begin the cleaning process.

Here are several methods that work well:

- Probably the least aggressive is to use a push broom (with bristles of medium stiffness) along with a non-film-forming detergent or a citrus-based cleaning solution. Wet the surface down with a garden hose, and then spray the cleaning solution over the surface and aggressively start to scrub in a circular motion. For large areas, the job goes faster if one person sprays the surface with water while several other workers do the scrubbing with the push brooms.

- A rotary floor buffer and a detergent scrub is another excellent way to remove the release. Some contractors like to use a bristle attachment while others prefer Nylo-Grit pads (which use a nylon designed for more aggressive scrubbing). Be careful when using pads, however, because they can leave tiny nylon fibers on the concrete surface that are hard to see—until you start putting the sealer down.

- Some contractors like to use a very diluted solution of muriatic acid to remove the release (about 10 to 15 parts water to 1 part muriatic acid). It's important to prewet the surface prior to applying the acid solution. Once the surface is wet, apply the acid with a sprayer or a garden-type sprinkling can and then scrub the surface with a broom. Only use plastic vessels to contain the acid solution, because it can react with metal. Muriatic acid is an excellent way to remove any stubborn spots you couldn't scrub off with a push broom or floor buffer. However, if you use a solution that's too strong, you could end up removing too much release and some of the color.

Regardless of the method you choose for release removal, it's a good idea when you're nearing the end of the job to step back and view the slab from different angles. You'll often find patches of release that will require more aggressive removal by spot scrubbing.

A rotary floor buffer fitted with a bristle attachment or nylon pad is also an effective release removal method.

When you're nearing the end of the job, look for stubborn patches of release that may require additional spot scrubbing.

Pressure washing is a fast way to remove residual release, but it can also be too aggressive and strip away some of the color. Proceed with caution, and make provisions to control water runoff.

After applying a patching material, take the same release agent used during stamping and lightly apply it to the affected area. Then use a texturing skin to gently pat the release into the patch.

CHAPTER 25

FIXING MINOR FLAWS

After release removal, you'll often discover a few spots that need touchup, or recoloring. You may also notice flaws that will require minor repair, such as plastic shrinkage cracks or footprint marks. Following are some tips for correcting these problems.

Recoloring

On almost all stamped concrete projects, there will be some subtle color imperfections. For example, maybe the pigmented release didn't take completely in one area or you removed too much release. Perhaps you had to chip away a bit of excess paste between the grout lines in the stamp imprint, exposing the underlying gray concrete. Or maybe you're just not satisfied with the color intensity in some spots. There are several ways you can restore or correct the color in these areas.

I've had good success using a mixture of solvent and clear solvent-based sealer tinted with the desired color of powdered release. I start with a ratio of two parts solvent (you can use zyxlene, lacquer thinner, or acetone) to one part sealer. Then I tint this mixture with the dry release and spray it onto the surface. The sealer locks in the color, and thinning the mixture with solvent breaks down the pigment particles so the concrete absorbs the color more readily.

If the area is small, you can simply moisten the surface with water and take the dry release and rub it in with a sponge. Some contractors prefer to use a liquid release and tint it with a powdered release. (This method is described in Chapter 13.)

Once you've applied these tints, you must lock in the color by applying more sealer.

Repairing minor cracks and other blemishes

Even if you take every measure possible to prevent plastic cracking, you may end with a few small plastic shrinkage cracks. For years, I used to get rid of the cracks by "erasing" them with a rubbing stone and a bit of water. The problem with this method is that it can erase some of the color as well and make the surface around the crack grainy in texture.

After some experimenting, I discovered a method that's much more effective. First, rub a small bit of water on the crack (an amount equivalent to dipping your finger in a glass of water). Then take a smooth-headed or ball-peen hammer and lightly tap the crack closed, approaching the crack from an angle rather than pinging the hammer straight down. It's amazing how well this method works.

After the surface has been cleaned, you may notice other flaws that require touchup. Sometimes the edges of the imprint will be frayed or chipped. This can happen during the stamping process if you allow the concrete to harden to such a degree that tamping of the mats blows out the edges. The edges can also fray if the stamping tool is not lifted evenly from the surface. If you used a dry-shake color hardener, you can make a patching material by mixing some of the dry hardener with a mixture of latex or acrylic bonding agent and water

Use a hand chisel to remove "squeeze" — displaced cement paste that comes up through the grout joints between stamps.

If a slight ridge of concrete remains after you remove the edge forms, use a broken piece of concrete as a rubbing stone to smooth it down.

(usually at a 1:1 ratio). After applying the patch with a margin trowel, take the same release agent you used during the stamping phase and lightly apply it to the affected area. Then use a texture skin to gently pat the release into the patch. If the concrete is integrally colored, color matching of the repair material will be more difficult. You will have to determine the proper dosage rate of color needed for the small amount of cement required for the patch. Consult with the manufacturer of the integral color for assistance with dosage rates.

Occasionally you'll find larger imperfections in the stamped concrete, such as an errant footprint. While some flaws can be successfully repaired, others are much harder to eradicate, especially if the problem area is too deep to ping or rub out. In these more extreme cases, what works well is to use a bush hammer to remove the entire problem area down to the lowest spot of the blemish. After bush hammering, clean away all debris and then premoisten the repair area. Mix equal parts of bonding agent and water, and then brush the solution onto the concrete. Before this solution dries, mix up a patching material combining the same color hardener used during the stamping phase with the bonding agent, and then trowel this mixture onto the surface. The consistency should be similar to cake icing. As described earlier, you can then use a texture skin or stamping tool to retexture the spot, along with the same release agent used originally.

Don't apply the patch material at a thickness exceeding 1/4 inch. Color hardener has a high cement content but contains no coarse aggregate, so you could get shrinkage cracking at greater thicknesses. If possible, cover the patched area with a piece of building paper or cardboard to slow down the rate of evaporation until the patch cures. This can minimize the amount of shrinkage cracking.

Using these techniques, you can disguise most imperfections. But keep in mind a patch is a patch. Practice makes perfect, and this especially true for the art of making repairs.

You can make plastic shrinkage cracks disappear quickly by lightly tapping them closed with a ball-peen hammer.

CHAPTER 26

SEALING STAMPED CONCRETE

No decorative stamped concrete installation is complete without the application of a sealer. This is the final step in the process and one of the most important. A sealer provides multiple benefits:

- It enriches the color intensity of the concrete, whether the color is integral, chemically stained, or obtained from a dry-shake hardener and antiquing release.

- It can add sheen to the surface ranging from satin to high gloss, depending on the product used.

- It greatly reduces the chance that efflorescence will discolor the surface (read the recommendations for combating efflorescence in Chapter 6).

- It blocks the penetration of stains from dirt, chemicals, leaves, and other substances, making the concrete easier to clean and maintain.

A general overview of sealers is provided here. Always check with the sealer manufacturer for recommendations as to the most appropriate sealer to use for a particular project as well as application guidelines.

Selecting a sealer

Although many types of sealers are available, the primary type used for exterior stamped concrete flatwork is a solvent- or water-based acrylic. For stamped interior floors, you may want to use a coating rather than a sealer (see Sealers versus Coatings on page 121).

Most acrylic sealers are easy to apply, economical, and breathable (they allow water vapor to escape from the concrete). They also adhere well to the surface even when applied shortly after final set.

I prefer solvent-based acrylics because they penetrate well and are less likely to

> **TIP**
>
> Sealed concrete flatwork can be slippery when wet. If slip resistance of the surface is a concern, such as on a pool deck or sloped pavement, additives are available that you can add to the sealer to increase the traction. At most specialty paint stores, you can purchase packages of grit in different gradations, depending on the level of skid-resistance needed. They provide a surface texture similar to fine sandpaper. Follow the recommended dosage rates given on the package. You can also add tiny glass or resin beads to the sealer to improve traction. Some sealer manufacturers sell products that already contain these additives.

What a difference a coat of sealer makes! Here, only half of the slab has been sealed. It's not hard to guess which half still needs the color enriching benefits a sealer can provide.

A sealer emphasizes the best qualities of stamped concrete by enriching the color intensity and making the surface easier to maintain.

> **TIP**
>
> **If you plan to grout the joints in the stamped pattern, you should seal the concrete first (as described in this chapter) to protect the stamping work from any grout residue, which could permanently stain the surface. Let the sealer dry overnight, and then come back and grind the sealer out of the joints using an angle grinder. This will ensure maximum adhesion of the grout to the concrete. I have good success applying the grout with a grout bag (similar to a cake-decorating bag) because it allows you to squeeze the grout directly into the joint with greater control. After cleaning up any grout residue and allowing the grouted joints to cure overnight, apply another coat of sealer to the entire surface.**

turn white, or milky—a potential problem when water-based sealers are applied too heavily or in moist conditions. Most solvent-based sealers provide a glossy sheen that enhances the colors in the stamped concrete. But if you don't want a shiny, wet look, these products also are available with low-gloss or matte finishes.

Some other factors to consider when selecting a sealer include:

- Durability
- Abrasion resistance
- UV stability (if used on exterior surfaces)
- Ease of application
- Cost

Preparing the surface

Proper surface preparation is critical to sealer performance. To obtain maximum adhesion and long-term durability, the sealer must be applied to a clean, dry surface.

Before applying the sealer, you should have already removed the release and cleaned the surface according to the recommendations given in Chapter 24. Also, the slab should have cured sufficiently. If you are using a cure and seal product (described in Chapter 22), the manufacturer will generally recommend waiting several weeks before applying the second coat (sometimes referred to as the seal coat, or finish coat).

When you go back to seal your project, be sure the surface is free of moisture, dirt, and residue. A leaf blower is a good tool to have on hand because you can use it to blow moisture out of joints or low spots in the pattern where water may have accumulated and to remove fallen leaves or other surface contaminants.

Applying the sealer

Following are some general guidelines for applying sealer. In most cases, though, you should adhere to the application instructions provided by the sealer manufacturer.

First, it's important to be aware of one of the main causes of sealer failure: applying it too thickly. Generally, the rate of application for most sealers ranges from 300 to 400 square feet per gallon for each coat. If you use a sealer with a higher percentage of solids (such as 30 percent or greater) and apply it too thickly, chances are it will just lie on the surface rather than penetrate into the surface pores. Sealers that "puddle" in this manner tend to turn white or milky. Typically, you'll get more uniform coverage and better performance by applying the sealer in two thin coats rather than one thick coat.

In my opinion, one of the best ways to apply sealer is with a pump-up metal sprayer commonly used for applying curing compounds. It atomizes the sealer so it goes on as a light mist. When using this method, I try to pump as much pressure into the sprayer as possible, which helps produce a finer mist.

Airless sprayers can also be used, but they are more likely to produce overspray, which is not only hazardous to inhale but can also settle on unprotected surfaces adjacent to the work area. You must be especially careful on windy days or if spraying indoors.

You can also apply sealer with a roller. Use a roller with a fairly thick nap (about 3/8 inch) to work the sealer into the textured surface and into depressions, such as grout lines.

One of the most effective techniques is to combine both spraying and rolling, especially when the stamped pattern has deep grout lines. The sealer tends to settle in these low-lying areas when applied by spray. Going back over the surface with a roller where necessary helps to distribute the sealer uniformly.

Use a roller with a fairly thick nap (about 3/8 inch) to work the sealer into the textured surface and pattern lines.

Sometimes in warm conditions or if the concrete surface is porous, bubbling in the sealer can occur. If you are left with bubbles after applying a solvent-based sealer, an easy fix is to spray or roll straight xylene onto the surface. The xylene re-emulsifies the sealer and gets rid of the bubbles. Xylene is extremely flammable, though, so use it with extreme caution.

Using a tinted sealer

Tinting your sealer is a great way to provide additional accenting (as discussed in Chapter 13, *Methods of Coloring Stamped Concrete*). But remember, if you're using a powdered release as the tint, it will only work when mixed into a solvent-based sealer since powdered releases repel water.

The best way to apply tinted sealer is with a pump-type sprayer. Make sure you continue to agitate the spray can by gently shaking it repeatedly during the spraying process so the tint stays in suspension. The tinted sealer settles into the nooks and crannies of the stamped pattern, where it leaves concentrated accents of color.

After applying an accent coat of tinted sealer, give it sufficient time to dry (between two to four hours), and then go over the top of it with a layer of clear sealer. This will help lock the tinted sealer layer into the surface.

The amount of tint to add to the sealer can vary, depending on the color intensity desired. A general formula to use as a starting point: For 5 gallons of tinted product, mix equal parts of sealer and xylene and add 1 cup of powdered release (you may need to add more or less release depending on the desired effect). The xylene helps dissolve the pigment particles in the release and thins the sealer, allowing the color to penetrate into the surface pores of the concrete.

Maintaining sealed concrete

Maintaining properly sealed, stamped concrete will depend on exposure conditions and the type and amount of traffic the surface will receive. Although a sealer will inhibit stains, the owner should still sweep and wash the surface occasionally to avoid dirt buildup. Heavily contaminated exterior surfaces can be pressure washed or scrubbed with a rotary floor scrubber and a mild detergent. For interior surfaces, wet mopping or dry dust mopping of the floor is normally the only regular maintenance needed. Also, for interior surfaces, application of a floor wax or polish can provide extra protection in high-traffic areas.

Over time, the sealer may begin to show some wear. If the surface begins to dull or lose its sheen, recoating with sealer can restore the luster. In high-traffic areas, such as shopping malls, it is especially important to maintain the sealed surface. Otherwise, wear patterns will begin to show, which can jeopardize the color of the release or other type of coloring, such as a stain.

Sealers versus Coatings

For interior stamped concrete, a coating such as an epoxy or aliphatic polyurethane may be preferable to a sealer. Here are the key differences: Sealers are generally low-viscosity, penetrating materials that fill the surface pores of the concrete substrate without leaving an impenetrable film that prevents the transmission of moisture. Coatings, on the other hand, are applied more thickly than sealers and are designed to build a surface film that provides better protection and easier cleaning of the substrate. Unlike acrylic sealers, most coatings do not permit moisture to escape from the slab.

To avoid trapping moisture in the slab when working with high-build coatings, most manufacturers recommend curing new concrete for a minimum of 28 days prior to coating application. This could prove to be a problem on jobs with an aggressive construction schedule. Regardless of the extended cure time, you should still conduct a moisture-vapor transmission test before applying coatings to concrete floor surfaces. The most effective way to conduct this test is with a calcium-chloride test kit. One source for these kits is Vaporprecision (see resources list on page 138.)

Try combining spraying and rolling when the stamped pattern has deep grout lines. Going back over the surface with a roller will evenly distribute any sealer that settles into crevices.

CHAPTER 27

TOOLS, EQUIPMENT, AND SUPPLIES

Up to this point, I've focused primarily on the materials and procedures you'll use to install decorative stamped concrete. But to perform each operation successfully, you must also use the proper tools and equipment for the task at hand.

If you're new to decorative stamping, this chapter presents the basic tools and equipment you'll need to get started, covering everything from grading and form setting to completing the stamping process. As your business expands and you begin tackling larger projects, you may need to add more heavy-duty equipment to your inventory.

To help you compile your shopping list, I've organized the supplies you'll need in four general categories:

- Grading, setting forms, and installing reinforcement
- Concrete placing and finishing
- Stamping and sealing
- Personal protection

Finishing and stamping tools come in a wide variety of shapes, sizes, materials, and costs. Check with your local supply house or attend training sessions to find out which tools are being used in the industry. Never cut corners, trying to get by with a tool you already own rather than purchasing the appropriate tool for the job. Using the right equipment will help you and your crew become more efficient while improving the quality of your work.

Grading, setting forms, and installing reinforcement
BASICS

Square-point shovels and concrete rakes (also called come-alongs)	To distribute fill for subgrade preparation, and to move fresh concrete ahead of the straightedge.
Vibratory plate compactor or rammer	For subgrade compaction.
Transit or laser level	To shoot elevations and determine slope.
String lines, chalk line reel and chalk	To set and mark forms for alignment and elevation. Chalk lines are useful for marking top-of-slab elevations on abutting buildings.
Formwork	Flatwork forms are made of various materials (including wood, plastic, and Masonite) and in various dimensions (ranging from 1x4 inches to 2x12 inches). See Chapter 15, *Erecting the Forms,* for additional guidance.
Stakes, braces, nails	To hold and secure the forms. Stakes are generally made of wood or steel. A duplex #8 is the most commonly used nail for fastening stakes and braces.
Tape measure, framing square	For taking measurements, measuring depth and to square forms.
Rebar cutters, tie wire, pliers, rebar supports	For cutting and installing reinforcing bars and wire mesh. Supports are needed to keep the reinforcement at the proper level in the slab (see Chapter 16, *Installing Reinforcement*).
Miscellaneous supplies	Marking pencil, nail apron or pouch, hammers, sledgehammers.

A skid steer is a multipurpose workhorse that can excavate, move fill, and haul fresh concrete.

BOBCAT

Equipment for subgrade compaction (from top clockwise): skid steer, tamper, vibratory plate compactor, roller.

WACKER CORPORATION

WACKER CORPORATION

WACKER CORPORATION

ADDITIONAL EQUIPMENT

Skid-steer loader or backhoe	To do excavating work and to move large quantities of fill and aggregate on commercial projects. A skid steer with a bucket can also be used on residential jobs to haul fresh concrete to the placement area (see Chapter 17, *Placing the Concrete*).
Roller compactor	For subgrade compaction on large jobs.

Concrete placing and finishing
BASICS

Strike off, or straightedge	To strike off the concrete surface and make it as smooth and level as possible before finishing. I prefer to use aluminum rather than wood straightedges (see Chapter 18, *Striking Off and Finishing the Concrete*).
Kneeboards, or sliders	Provide a flat pad for finishers to kneel on when it's necessary to work out on the slab, such as to hand finish around columns or up against a wall. Made from a variety of materials including plywood, melamine, foam, and aluminum.
Margin trowel (also called a pointer or pointed mason's trowel)	A must for every stamper's toolbox. It has multiple uses, including scraping off concrete from finishing tools, knocking aggregate away from the forms, mixing materials, and patching.
Bull float	One of the most important tools for preparing the slab for stamping. Serves a multitude of functions. Use to eliminate high and low spots, embed large aggregate at the surface, bring a layer of paste to the surface needed during final finishing, and float in dry-shake color hardener. Bull float blades are typically made of wood, magnesium, resin, or composite materials; lengths are generally 3 to 4 feet. Magnesium floats are best for finishing air-entrained concrete (as noted in Chapter 18). Long handles either clip on or screw into the float head so it can be pushed out onto the slab while the user stands at the perimeter.
Hand float	A smaller handheld version of the bull float, ranging in length from 12 to 18 inches. Especially useful for floating along the perimeter of the forms. Materials include wood, magnesium, and resin.
Darby	Basically a longer version of a hand float, ranging in length from 2 to 4 feet. Useful for leveling problem areas.
Hand trowel	A thin, flat steel blade used after floating to compact the paste layer at the surface and provide a smooth, flat finish. Available in different shapes (with rounded or square edges) and lengths (ranging from 8 to 24 inches). Smaller trowels are useful for borders, work in restricted areas, or to work in flash accents of color hardener.
Fresno	A large trowel (about 2 to 4 feet in length) used for final finishing after bull floating (and after bleed water has evaporated from the surface). Like hand trowels, fresnos come with round or square edges. Rounded-edge types are less likely to leave trowel lines. Long handles (like those used for bull floats) either clip on or screw into the blade.
Hand edger	To edge along construction or isolation joints and along the perimeter of the forms. Proper edging provides a finished radius while compacting and densifying the edge, making it less prone to chipping. The result is a much cleaner look. Edgers come in different radii ranging from about 1/8 to 1 inch. The most commonly used are 1/4, 3/8, and 1/2 inch.
Groover (also called a deep jointer)	A hand-held tool with a V-shaped bit used to tool contraction joints in the plastic concrete (as an alternative to sawcutting). The bits are usually made of bronze or stainless steel and come in various depths. Use a bit with a depth at least one-fourth the slab thickness (see Chapter 23, *Installing Joints*). A helpful hint: Use a 1x12 inch board to groove up against. It will distribute your weight more evenly than a 2x4 so you won't leave a depression in the fresh concrete. You can also purchase groovers with handle attachments.
Walking tools (edgers, jointers, combination tools)	Equipped with long handles to allow grooving or edging of concrete from a standing rather than kneeling position. A combination tool is designed to both edge and groove. These tools are timesavers on large slabs with long stretches of formwork.
Scrub brush	To clean concrete from finishing tools. Never use your hands; extended use of trowels produces extremely sharp edges.

ADDITIONAL EQUIPMENT

Small-line pump or power buggy	To convey concrete to restricted jobsites (see Chapter 17).
Vibrating screed	To strike off low-slump concrete efficiently.
Cut-off saw with a diamond or abrasive blade	To cut contraction joints in the slab.
Water hose	To clean the surface and wet the subgrade if needed.

Assorted handheld finishing tools

Edger

Magnesium float

Resin float

A tampico brush, to apply powdered releases.

Miscellaneous supplies for applying sealer and cleaning the slab.

Stamping and sealing

BASICS

Full set of stamping mats	Start with three or so basic patterns, such as a slate, cobblestone, and brick. Rigid and semi-rigid mat types are available, each having positive attributes (for more guidance on choosing stamps, see Chapter 21, *The Concrete Stamping Process*).
Flex mat (or floppy)	Very flexible, bendable mats for stamping up against vertical surfaces, such as columns and walls. Get flex mats that match the pattern and texture of your rigid stamping tools.
Texture skins	Even thinner and more pliable than flex mats with no grout joints, just texture. Used to texture slab perimeters and vertical faces, such as stair risers. Can also be used to fix blemishes from nonuniform stamping and instead of rigid stamping tools, when only texture is desired.
Tamper	To press the stamps firmly into the concrete surface, ensuring a complete imprint. The type of tamper you use will depend on the mat type. Metal dirt tampers (available at most hardware stores) work well with semi-rigid or rigid mats. For thinner texture skins, use a wider polyurethane tamper (available from stamping tool suppliers). The pressure it exerts is less concentrated, so the outline of the tamper won't transfer to the concrete surface.
Tampico brush	To apply powdered releases.
Pump-up or airless sprayer	To apply sealer and liquid release agents.
Roller frames with roller heads	To apply sealer.
Hand rollers or chisels	To detail joints.
Miscellaneous supplies	Rollers (with 3/8 inch nap for applying sealer), citrus-based cleaner (for removing powdered releases), muriatic acid (for stubborn spot removal).

ADDITIONAL SUPPLIES

More stamp sets	As your business grows, add additional patterns to your stamping tool collection so you can offer customers greater variety.
Leaf blower	To get rid of standing moisture and blow off any contaminants on the slab before applying sealer.

A hand roller for detailing joints.

Personal protection
Review Chapter 11, *The Importance of Safety*

Nonabsorbent rubber boots, rubber or latex gloves	Wear when working with fresh concrete to avoid contact with the cement in the concrete. Also for handling color hardener and releases.
Eye protection (approved safety glasses or goggles)	To keep concrete splatters, dust, and other construction debris out of your eyes.
Hard hat	Required head protection on most commercial construction projects.
Dust mask or respirator	To protect from inhalation of dry-shake hardeners, powdered release agents, and dust from construction operations (such as sawcutting).
Kneepads	To cushion your knees when finishing concrete.
Ear protection	Wear when operating or working in the vicinity of loud equipment, such as power saws.

CHAPTER 28

HOW TO SELL STAMPED CONCRETE WORK

Even the finest, most experienced craftsman needs to get the word out about his business. Some contractors who produce decorative stamped concrete believe the work will sell itself and no energy should be spent promoting their business. More often than not, this is just not true. In fact, sometimes contractors with lesser ability will go farther in business because they promote and market their services more effectively.

Listed in this chapter are what I consider the fundamentals to selling stamped concrete.

Showcase phenomenal work. Each job should be looked at as a calling card for your company. We discussed the importance of chronicling your work in Chapter 10. One of the main purposes of chronicling is to build a portfolio that showcases your artistry. Your successful stamped concrete installations can be your most powerful sales tool.

Open a showroom. Color charts and professional samples are also an essential part of your sales toolkit. If you have the budget and space, consider opening a showroom or design center to display samples of the stamping patterns and color selections you offer (refer back to Chapter 8 for tips on preparing samples).

The showroom doesn't have to be indoors or limited to a small array of samples. A colleague of mine in Philadelphia has done his whole parking lot in stamped concrete, using different color schemes, textured borders, and various stamp patterns. That's probably his best sales tool. He can invite designers and homeowners to look at virtually every pattern and color he has to offer.

Establish a professional web site. Today, many buyers of products and services turn to the Internet first for information. Be sure to establish an online presence so your firm isn't overlooked. A web site introduces

A design center displaying stamp patterns and color selections provides a stimulating atmosphere where clients can explore the many possibilities.

COLORADO HARDSCAPES

new prospects to your company and helps them learn more about your background and capabilities, without the pressure of a sales call. A web site can also save you time by allowing prospective customers to review your stamp patterns, colors, designs, and portfolio before you even meet them in person.

Continually hone your craft. In the stamped concrete business, there is always more to learn no matter what your skill level. Attend training seminars to master new techniques or to refine your existing skills. Network with others in the stamped concrete industry at events such as the World of Concrete and gatherings of the Decorative Concrete Council. Stay abreast of the latest stamping tools, coloring agents, sealers, and other products for stamped concrete.

Concentrate your marketing efforts. Use the rifle approach rather than the shotgun approach to selling. General contractors, architects, builders, and designers will award you repeat work if you do a great job for them. By concentrating on building strong relationships with these core clients, you are likely to earn most of their business. You will also be able to count on these clients to give your company excellent references.

Utilize your supplier network. Your suppliers can also be good sources of referrals. This group includes your ready-mix producer, stamping tool supplier, color hardener supplier, and even your rental equipment dealer. Keep them all informed of interesting projects you've done—particularly jobs where you have used their materials or equipment. Drop off pictures of your work occasionally so they can see what you are capable of doing.

Deliver excellent customer service. Earning a reputation for reliability, as well as excellent work, will serve you well. Return phone calls, fix your mistakes, and dress professionally. Be proactive and address any concerns before they escalate into bigger problems. Being attentive to your customer's needs can set your firm apart from the competition.

Contribute design ideas. Rarely are specifications for stamped concrete clear-cut, even on commercial projects. In most cases, owners, designers, builders, and architects will welcome your design ideas and suggestions.

Having an artistic vision will work to your advantage. If you lack inspiration, however, revisit Chapter 3 for some good sources of design ideas.

Be honest and straightforward. Tell prospects about all of the wonderful aspects of stamped concrete, but also don't neglect to tell them about the variables (as discussed in Chapter 8). Your honesty will be appreciated, and you will find that most people are fine with the realities of stamped concrete—as long as they know what to expect.

Be patient. Becoming proficient at stamped concrete does not happen overnight. Likewise, building a business on a solid foundation takes time. Decide early on if you have the motivation and desire to build a business the right way—slow and steady.

It takes many years to become an overnight success!

Be attentive to your customers' needs. Offer design ideas, explain the stamping process, and patiently address all questions and concerns.

Create a frame for your work by installing borders in contrasting colors.

CHAPTER 29

TAKING STAMPED CONCRETE OVER THE TOP

Once you develop the basic skills necessary to consistently produce quality stamped concrete, consider taking your work to new levels of originality.

Here are some ways to distinguish your stamped concrete installations from those of your competitors:

Add a border. A border can serve as a decorative frame for your artistry, whether used as a subtle accent or for bold contrast. You can stamp the borders in a contrasting pattern (such as a brick border framing a flagstone-patterned walkway) or apply a different type of architectural concrete finish, as discussed later. For a totally unique look, use custom stamps to create border designs that replicate nature, such as fern leaves, seashells, or fossil renderings.

Incorporate other architectural finishes. Enhance or complement the colors and textures of your stamped designs by incorporating other architectural finishes, either in a border, as an inset, or in bands or fields separated by joints. Some alternatives include light sandblasting (some manufacturers offer templates for producing custom designs), using surface retarders to expose the aggregate in select areas, or providing contrast with a simple broom finish.

Combine stamped concrete with other materials. You can also combine stamping with other types of materials to add diversity. Some options include brick pavers, natural stone, or redwood divider strips.

Mix it up. Incorporate several different stamp patterns into your designs.

Work with a multitude of colors. Expand your color palette by combining a variety of coloring methods, such as integral colors, dry-shake hardeners, chemical stains, dyes, and tints (see Chapter 13). Don't limit yourself to one or two hues. With dry-shake hardeners, for example, try combining multiple colors, using one as a base shade and the others as flash accents. These techniques are described in Chapter 19.

Go vertical. Decorative stamping doesn't have to be limited to horizontal concrete surfaces, such as patios and driveways. You can use a similar process on vertical surfaces. Outdoors, echo the pattern from a stamped patio or pool deck on the surfaces of surrounding privacy walls. Indoors, use vertical stamping to turn a wall, fireplace front, or backsplash into a dramatic focal point. These effects are possible with new lightweight, cementitious, polymer-modified overlay mixes that can be applied by trowel directly to drywall, masonry, existing concrete, and other vertical surfaces. Because they are lightweight, these wall mixes can go on at thicknesses of up to 3 inches without sagging. While the overlay is still workable, custom wall stamps are used to produce deep-relief patterns and textures that mimic hand-laid stone or rock. Finishing the wall with earth-toned stains completes the transformation. You can find suppliers of vertical stamping systems in the list of resources on page 138.

These are just a few examples of ways you can go from ordinary to extraordinary with stamped concrete. Use your imagination to create your own signature looks.

Mimic materials other than stone or brick, such as wood planking.

Dazzle with a rainbow of colors.

Cover the walls. Why keep stamped concrete underfoot?

Add drama with insets or fields of plain concrete.

Use flash accents to create subtle, natural-looking color variations.

GLOSSARY

A

abrasion resistance – How well a concrete surface resists being worn away by friction or rubbing.

absorption – The process by which a liquid is drawn into and tends to fill permeable pores in a porous material, such as concrete.

accelerator – An admixture used to shorten the set time of concrete and/or speed strength development.

admixture – An ingredient in concrete—other than water, portland cement, and aggregate—used to modify the properties of concrete in its freshly mixed, setting, or hardened states.

aggregate – A granular material such as sand, rock, crushed stone, gravel, or other particles added to concrete to improve its structural performance.

air content – The amount of entrained or entrapped air in concrete, usually expressed as a percentage of total volume.

air-entraining admixture – Added to fresh concrete to cause the development of a system of microscopic air bubbles. Helps to improve the freeze-thaw resistance of hardened concrete.

B

blast furnace slag – A glassy, granular material formed when molten blast furnace slag is rapidly chilled. Ground granulated slags are sometimes used in concrete mixtures as a partial replacement for portland cement to help reduce permeability and improve durability.

bleeding, bleed water – A form of segregation in which some of the water in a mix rises to the surface of freshly placed concrete.

bond – Degree of adhesion of the cement paste to aggregate, reinforcement, or other materials. Also, adhesion of a sealer or other material applied to the concrete surface.

bond breaker – A material used to prevent adhesion of newly placed concrete to other materials.

broom finish – Surface texture obtained by pushing a broom over freshly placed concrete.

bull float – A tool with a large, flat, rectangular blade usually made of wood, aluminum, or magnesium. Used in finishing operations to smooth the surface of freshly placed concrete while leaving a relatively open texture.

bush hammer – A percussive hammer with rows of pyramid-shaped points used for removing the surface cement.

C

cast in place – Concrete placed and finished in its final location.

cement (portland) – A hydraulic product that sets and hardens when it chemically interacts with water. Made by burning a mixture of limestone and clay or similar materials.

cementitious materials – Substances that have hydraulic cementing properties (set and harden in the presence of water). Includes such materials as blast furnace slag, pozzolans, fly ash, natural cement, and hydraulic hydrated lime.

coarse aggregate – Graded granular material with a nominal maximum size ranging from 3/8 inch to 1 1/2 inches.

cold joint – A visible delineation that forms in a concrete element when the placement of concrete is delayed. The concrete in place hardens prior to the next placement of concrete against it.

color hardener – A mixture of coloring pigments, cement, and surface conditioning agents. Applied as a dry-shake to fresh concrete to produce a colorful, wear-resistant surface.

compressive strength – The maximum compressive stress concrete is capable of sustaining, expressed as pounds per square inch (psi).

concrete – A composite material consisting of a binding medium within which aggregate particles are embedded. In portland cement concrete, the binder is a mixture of portland cement and water; the filler may be any of a wide variety of natural or artificial fine and coarse aggregates.

consistency – The ability of fresh concrete to flow. The usual measure of consistency is slump.

crusting – A condition that occurs when the surface of freshly placed concrete dries too quickly, often due to exposure to direct sun, wind, or high temperatures.

curing – Action taken to maintain moisture and temperature conditions of freshly placed concrete during a defined period of time following placement. Helps to ensure adequate hydration of the cementitious materials and proper hardening of the concrete.

curing compound – A liquid that, when applied to the surface of newly placed concrete, forms a membrane that retards the evaporation of water.

dry-shake method – A procedure used to hand broadcast dry color hardener thinly and evenly across the concrete surface, usually in a continuous motion.

drying shrinkage – A decrease in the volume of concrete as it dries.

durability – The ability of concrete to resist weathering exposure, chemical attack, and abrasion.

edger – A tool used on the edges of fresh concrete to provide a clean, finished edge.

efflorescence – A crystalline deposit of salts (usually white in color) that forms on the concrete surface when soluble calcium hydroxides leach from the concrete and combine with carbon dioxide in the atmosphere.

exposed-aggregate concrete – A decorative surface formed by washing away cement paste from the concrete surface to expose the underlying aggregates.

fine aggregate – A graded granular material entirely passing the 3/8 inch sieve.

finishing – Leveling, smoothing, compacting, and otherwise treating the surface of newly placed concrete to produce the desired appearance and service properties.

flashing (or flash accent) – A technique for applying accent colors of hardener to concrete surfaces before stamping to achieve contrast. Results in subtle, natural-looking color variations.

floating – The next-to-last stage in concrete finishing (preceding troweling) when a bull float or hand float is used to smooth the concrete and bring water to the surface.

fly ash – A byproduct resulting from the combustion of ground or powdered coal; sometimes used as a cementitious material in concrete.

form (or formwork) – A temporary structure or mold to support and contain concrete while it is setting and gaining sufficient strength to be self-supporting.

grade line – The top of a concrete placement established using a string line.

GLOSSARY

groover – A tool with a V-shaped bit used to create contraction joints in plastic concrete.

H

hard-troweled finish – Surface finish obtained by using a trowel with a steel blade for final finishing of concrete. Often used where a smooth, hard, flat surface is desired.

hydration – The chemical reaction between cement and water that causes concrete to harden.

I

integral color – A coloring agent usually premixed into the concrete at the batch plant. Can be used instead of or with color hardener, but the colors are usually less vibrant.

J

joint (contraction, expansion, or isolation) – Formed, sawed, or tooled groove in a concrete slab used to regulate the location of cracking (contraction joint) or to allow expansion or movement of adjoining structures.

K

kneeboards – Boards used by concrete finishers to kneel on when hand floating or troweling concrete flatwork.

M

mix design – Specific proportions of ingredients (cement, aggregates, water, and admixtures) used to produce concrete suited for a particular set of job conditions.

monolithic placement – Concrete cast as a single, one piece, integral structure.

P

plastic – A condition of freshly mixed concrete indicating that it is workable and readily moldable, is cohesive, and has an ample content of fines and cement but is not overly wet.

plastic shrinkage cracks – Short, irregular cracks that can develop in concrete within the first 24 hours after placement.

plasticity – Property of freshly mixed concrete, cement paste or mortar which determines its ease of molding or resistance to deformation.

popout – A pit or crater in the concrete surface, ranging in size from 1/4 inch to several inches in diameter, that results from the fracturing of unsound aggregate particles due to expansion pressure. Usually caused by porous aggregate having a high rate of absorption.

pozzolan – A siliceous or siliceous and aluminous material that, in the presence of moisture, chemically reacts with calcium hydroxide to form compounds possessing cementitious properties.

R

ready-mixed concrete – Concrete that is batched or mixed at a central plant before delivery to the jobsite for placement.

rebar (or reinforcing bars) – Ribbed steel bars installed in cast-in-place concrete construction to strengthen the concrete. Rebar come in various diameters and strength grades.

rebar spacing – The distance between parallel reinforcing bars, measured center to center.

rebar support – A rigid device used to support or hold reinforcing bars in proper position to prevent displacement before or during concrete placement.

reentrant corner – An angle in a concrete slab that points inward. Often vulnerable to cracking, unless a control joint is installed.

reinforced concrete – Concrete construction that has steel rebar or welded wire mesh embedded in it to provide greater tolerance to tension and stress.

release agent – A parting agent applied to the concrete surface and texturing mats before stamping to keep the mats from sticking to the plastic concrete. Also prolongs the life of the mats by decreasing the friction between the mat and the concrete.

scaling – The flaking or breaking away of a hardened concrete surface, often due to exposure to freezing and thawing.

sealer – Solvent- or liquid-based material used to protect and enhance the appearance of decorative stamped concrete.

segregation – The separation of the components of wet concrete caused by excessive handling or vibration.

set – The condition reached by concrete when plasticity is lost, usually measured in terms of resistance to penetration or deformation. Initial set refers to concrete that has reached first stiffening. Final set occurs when concrete attains full rigidity.

setting – The chemical reaction that occurs after the addition of water to a cementitious mixture, resulting in a gradual development of rigidity.

slip dowels – Plain, round steel bars that extend from one concrete placement into the next. They are used as a load-transfer device and to increase strength in the joint.

slope – The incline angle of a concrete surface, as a ratio of the rise (in inches) to the run (in feet).

slump – A measure of consistency of freshly mixed concrete, as determined by the distance the concrete "slumps" after a molded specimen is removed from an inverted funnel-shaped cone.

straight edge – A rigid, straight piece of wood or metal used to strike off a concrete surface to proper grade before the floating operation.

strike off – To level off freshly placed concrete to the correct elevation.

tamper – A handheld impact tool used to firmly press texturing mats into the plastic concrete.

texturing mats – Rigid or semi-flexible polyurethane tools for imprinting stone, slate, brick, and other patterns in concrete surfaces.

texturing skins – Flexible skins for adding seamless textures to concrete surfaces. Generally thinner and more pliable than mats.

trowel – A thin, flat steel tool, either pointed or rectangular, used to give concrete surfaces a dense, smooth finish.

water-cement ratio – The ratio of the amount of water to the amount of cement in a concrete mixture.

water reducer – An admixture that either increases the slump of freshly mixed concrete without increasing water content or maintains workability with a reduced amount of water.

welded wire mesh – A woven mesh of wire strands, welded at each intersection, used to reinforce concrete slabs. Also called welded wire fabric.

wet screeds – Strips of concrete placed at the proper elevation to act as height guides when pouring a concrete slab.

workability – The ease of which freshly mixed concrete can be mixed, placed, compacted, and finished.

RESOURCES

Ready to get started with decorative concrete stamping? Here are some good resources for tools, supplies, equipment, educational materials, and training. Many suppliers will ship their products worldwide.

AMERICAN CONCRETE INSTITUTE
38800 Country Club Drive
Farmington Hills, MI 48331
Phone: 248-848-3700
Fax: 248-848-3701
www.concrete.org

Produces more than 400 technical documents, reports, guides, specifications, and codes for the best use of concrete.

BOBCAT
250 E. Beaton Dr.
P.O. Box 6000
West Fargo, ND 58078
Phone: 701-241-8700
www.bobcat.com

A source for material handling equipment, such as skid-steer loaders, backhoes, excavators, and utility vehicles.

BOMANITE CORPORATION
232 S. Schnoor Ave.
Madera, CA 93637
Phone: 559-673-2411
Fax: 559-673-8246
www.bomanite.com

Produces dry-shake color hardener, integral color, release agents, imprinting tools, fibers, and other products for decorative concrete paving.

BRICKFORM RAFCO PRODUCTS
11061 Jersey Blvd.
Rancho Cucamonga, CA 91730
Phone: 800-483-9628
Fax: 909-484-3318
www.brickform.com

Carries a full line of decorative concrete products including mats, sandblast stencils, acid stains, integral color, dry-shake color hardener, overlays, and sealers.

COBBLECRETE INTERNATIONAL
1876 North 2700 West #1
Provo, Utah 84601
Phone: 888-224-6662
Fax: 801-225-1690
www.cobblecrete.com

Sells urethane stamps, release agents, color hardeners, iron-oxide pigments, acid stains, and water- and solvent-based sealers. Products can be shipped worldwide.

CONCRETE IMPRESSIONS
P.O. Box 34406
San Antonio, TX 78265
Phone: 210-646-8500
Fax: 210-646-0556
www.concreteimpressions.com

Carries a wide array of imprinting tools including mats, skins, rigid polyurethane stamps, and aluminum stamps. Also specializes in coloring systems for concrete, including acid stains, integral color, color hardener, and water-based tints.

CONCRETENETWORK.COM, INC.
11375 Oak Hill Lane
Yucaipa, CA 92399
Phone: 866-380-7754
Fax: 909-389-7744
www.concretenetwork.com

The Concrete Network provides a window to the world of concrete products, concrete services, and concrete service providers. Visitors to the site can find information on many popular concrete topics, including decorative concrete floors, concrete countertops, decorative concrete pool decks, patios, driveways, and much more.

CONCRETE SOLUTIONS, INC.
3904 Riley St.
San Diego, CA 92110
Phone: 800-232-8311
Fax: 619-297-3333
www.concretesolutions.com

Supplies products for the repair, restoration, and beautification of existing surfaces, such as a stampable overlay system, a spray-applied polymer-modified cement for recoloring or restoring concrete, and a decorative color-flake system for producing granite or terrazzo looks.

CONCRETE TEXTURING TOOL & SUPPLY
45 Underwood Road
Throop, PA 18512
Phone: 570-489-6025
Fax: 570-383-6711
www.concrete-texturing.com

A distributor of texture mats, color hardeners, and antique release agents, with 40 different in-stock colors and custom color matching available. Also carries integral color additives, sealers, and epoxies.

DAVIS COLORS
3700 East Olympic Blvd.
Los Angeles, CA 90023
Phone: 800-356-4848
Fax: 323-269-1053
www.daviscolors.com

Offers a wide spectrum of integral colors for use in ready-mixed or stamped concrete.

DECORATIVE CONCRETE COUNCIL/ AMERICAN SOCIETY OF CONCRETE CONTRACTORS
2025 S. Brentwood Blvd.
St. Louis, MO 63144
Phone: 314-962-0210
Fax: 314-968-4367
www.ascconline.org

A committee of decorative concrete contractors with the mission to advance the quality and use of decorative concrete systems.

DECORATIVE CONCRETE FINISHES, INC.
118 Pearl Industrial Ave.
Hoschton, GA 30548
Phone: 888-379-3779
Fax: 706-654-5466
www.dcf-usa.com

Manufacturer and supplier of specialty concrete products used for stamped concrete, polymer overlay stamping, chemical staining, and concrete sealing.

DECORATIVE CONCRETE INSTITUTE
8729 South Flat Rock Road
Douglasville, GA 30134
Phone: 877-324-8080
Fax: 770-489-4948
www.decorativeconcreteinstitute.com

Bob Harris' Decorative Concrete Institute provides consulting, education, installation, and on-the-job training to architects, artists, concrete finishers, faux finishers, general contractors, and interior designers across the U.S. and internationally. Some of the topics covered in the curriculum include decorative stamping, staining techniques, faux finishes, stenciling, design layout, decorative score cutting, and sandblasted and engraved graphics.

FLEX-C-MENT LLC
1810-1 East Poinsett St.
Greer, SC 29651
Phone: 864-877-3111
www.flex-c-ment.com

Specializes in cementitious overlay materials for producing stamped concrete floors, walls, and countertops. Also sells reusable rubber stamps for producing deep-relief stone or masonry wall textures.

FOSSILCRETE
121 NE 40th St.
Oklahoma City, OK 73105
Phone: 405-525-3722
Fax: 405-525-3367
www.fossilcrete.com

Offers a unique assortment of concrete stamping tools replicating the art forms of nature. Categories include animals and tracks, fossils, plants and trees, and marine life. Also supplies a specialized mix for stamping of vertical surfaces.

FRITZ-PAK CORPORATION
11220 Grader St. Ste. 600
Dallas, TX 75238
Phone: 888-746-4116
Fax: 214-349-3182
www.fritzpak.com

Sells premeasured, powdered admixtures packaged in ready-to-use water-soluble bags. Product line includes set retarders, superplasticizers, water reducers, accelerators, and air entrainers.

INTERNATIONAL SURFACE PREPARATION
6040 Osborn St.
Houston, TX 77054
Phone: 800-374-4043
Fax: 713-644-1785
www.surfacepreparation.com

Worldwide distributor of concrete cutting, grinding, and surface preparation equipment. The source for the Crac-Vac for decorative straight cutting, handheld and walk-behind grinders for profiling and mastic removal. Also supplies shotblasting equipment for surface preparation and profiling.

KRAFT TOOL COMPANY
8325 Hedge Lane Terrace
Shawnee, KS 66227
Phone: 913-422-4848
Fax: 913-422-1018
www.krafttool.com

Offers a full line of concrete leveling and finishing tools, including spreaders, straightedges, bull floats, steel trowels, hand floats, edgers and groovers, and kneeboards. Also supplies texture mats, color hardener, and powdered release agent.

L. M. SCOFIELD COMPANY
6533 Bandini Blvd.
Los Angeles, CA 90040
Phone: 800-800-9900
Fax: 323-720-3030
www.scofield.com

Provides engineered systems for coloring, texturing, and improving the performance of architectural concrete. Coloring admixtures, floor hardeners, colored cementitious toppings, stains, curing agents, sealers, coatings, repair products, and texturing tools are among its products.

RESOURCES *continued*

MARSHALLTOWN COMPANY
104 South 8th Ave.
Marshalltown, IA 50158
Phone: 641-753-5999
Fax: 641-753-6341
www.marshalltown.com

Manufacturer of finishing tools for concrete, including bull floats, trowels, hand floats, edgers and groovers, kneeboards, chisels, tampers, and more.

MATCRETE
Phone: 877-662-8273
Fax: 714-979-5478
www.matcrete.com

Carries over 300 patterns of stamping tools, including cast-aluminum tools and texturing mats. Also supplies color hardeners, release agents, sealers, and hand tools for touching up joints. Ships products worldwide.

PORTLAND CEMENT ASSOCIATION
5420 Old Orchard Road
Skokie, IL 60077
Phone: 847-966-6200
Fax: 847-966-8389
www.cement.org

A resource for technical documents on architectural and decorative concrete.

PROFESSIONAL DECORATIVE CONCRETE SERVICES/ RENEW-CRETE SYSTEMS
6869 Stapoint Court, Suite 115
Winter Park FL 32792
Phone: 866-531-9779
Fax: 407-677-7978
www.renewcrete.com

Sells many patterns of polyurethane mats and texturing skins as well as aluminum stamping tools. Other products include liquid and powdered release agents, dry-shake color hardener, integral color, and sealers.

PROLINE CONCRETE TOOLS
Phone: 800 221-9469
Fax: 702-873-7599
www.prolinestamps.com

Supplier of stamping tools and seamless texturing skins in stone, brick, slate, tile, and wood patterns. Ships worldwide.

QC CONSTRUCTION PRODUCTS
232 South Schnoor Ave.
Madera, CA 93637
Phone: 800-453-8213
www.qcconprod.com

Has a wide range of colorants for achieving hues ranging from subtle to bright, such as integral coloring, patina stain, dry-shake color hardener, and a penetrating water-based tinting compound. Also sells protective sealers and stripping products for removing worn sealer and finishes.

ROTEC INTERNATIONAL
3212 N. 40th St., Unit 400
Tampa, FL 33605
Phone: 813-969-4141
Fax: 813-968-9860
www.rotecsite.com

Manufacturer of chemicals and stamping tools for decorative concrete. Products include color hardener, sealer, powdered release, mats, and texturing skins.

SCHWING AMERICA INC.
5900 Centerville Road
St. Paul, MN 55127
Phone: 651-429-0999
Fax: 651-429-3464
www.schwing.com

Manufactures a broad line of concrete pumps, including trailer-mounted small-line models.

SOLOMON COLORS
P.O. Box 8288
Springfield, IL 62791
Phone: 800-624-0261
Fax: 217-522-3145
www.solomoncolors.com

Produces dry and liquid pigments for ready-mix concrete. Other products include decorative stamps, dry-shake color hardener, colored release, and clear liquid release.

SPECIALTY CONCRETE PRODUCTS INC.
1327 Lake Dogwood Dr.
West Columbia, SC 29170
Phone: 800-533-4702
Fax: 803-955-0011
www.scpusa.com

Manufacturer of stamped concrete products, acid stains, sealers, integral colors, and resurfacing materials for new concrete construction or renovation. Also offers hands-on decorative concrete training classes monthly at its South Carolina facility.

THE STAMP STORE
121 NE 40th St.
Oklahoma City, OK 73105
Phone: 888-848-0059
Fax: 405-525-3367
www.thestampstore.com

Offers an extensive line of texturing mats and skins in hundreds of stone, tile, slate, brick, and wood patterns. Also supplies stamped overlay mixes, integral colors, chemical stains, color hardeners, release agents, and stencils.

STAMPMASTER/CREATIVE URETHANE CONCEPTS INC.
907 Garland St.
Columbia, SC 29201
Phone: 888-901-6287
Fax: 803-376-4528
www.stampmaster.net

Offers more than 100 patterns and textures of urethane stamps, including natural stone, slate, tile, brick, and wood.

SUPER-KRETE INTERNATIONAL
1290 N. Johnson Ave., Suite 101
El Cajon, CA 92020
Phone: 800-995-1716
Fax: 619-401-8288
www.super-krete.com

Supplies a stampable overlay system for use on existing concrete surfaces, available in 16 colors.

SUPERSTONE INC.
1251 Burlington St.
Opa-Locka, FL 33054
Phone: 800-456-3561
Fax: 305-681-5106
www.superstone.com

Supplies a broad range of products for concrete resurfacing and beautification, including texturing mats, penetrating chemical stains, liquid colorant, integral color, solvent-based acrylic sealers, and resurfacing and crack repair polymers.

VAPORPRECISION
2941 West MacArthur Blvd., Suite 135
Santa Ana, CA 92704
Phone: 800-449-6194
Fax: 714-549-8245
www.vaportest.com

Supplies calcium chloride kits, tools, and technical support for moisture vapor testing of interior concrete slabs.

WACKER CORPORATION
N92 W15000 Anthony Ave.
P.O. Box 9007
Menomonee Falls, WI 53051
Phone: 800-770-0957
Fax: 800-822-0710
www.wackergroup.com

A source for power buggies, soil compaction equipment, concrete finishing machines, cut-off saws, and demolition equipment.

WAYNE DAVIS CONCRETE CO.
10 Wayne Davis Dr.
Tallapoosa, GA 30176
Phone: 770-574-2326
Fax: 770-574-7605
www.waynedavisconcrete.com

Ready-mix supplier with eight plants in Georgia.

SPECIAL THANK YOU FOR PHOTOS PROVIDED BY THE FOLLOWING CONTRACTORS:

AMCON, INC.
10 Park Ave.
Gaithersburg, MD 20877
Phone: 301-924-4910
www.amconcrete.com

COLORADO HARDSCAPES
8085 E. Harvard Ave.
Denver, CO 80231
Phone: 303-750-8200
www.coloradohardscapes.com

DECOSUP INC.
8232 NW 56 St.
Miami, FL 33166
Phone: 800-788-0014
www.decosup.com

DISTINCTIVE CONCRETE
P.O. Box 325
Rowley, MA 01969
Phone: 978-948-2970
www.distinctiveconcrete.com

L.L. GEANS CONSTRUCTION CO.
1923 N. Home St.
Mishawaka, IN 46545
Phone: 574-255-9671
www.llgeans.com

SULLIVAN CONCRETE TEXTURES
1111 Baker St.
Costa Mesa, CA 92626
Phone: 800-447-8559
www.sullivanconcrete.com

VERLENNICH MASONRY AND CONCRETE
1406 Prairie Ave. Unit D
Staples, MN 56479
Phone: 218-894-0074
www.stampedinstone.com